REDEFINING DIVERSITY AND DYNAMICS OF NATURAL RESOURCES MANAGEMENT IN ASIA

VOLUME 4

Reciprocal Relationship between Governance of Natural Resources and Socio-Ecological Systems Dynamics in West Sumatra, Indonesia
Rudi Febriamansyah, Yonariza, Raza Ullah and Ganesh P. Shivakoti; editors

Volume 4 is dedicated to Nobel Laureate Elinor Ostrom who is the source of inspiration in drafting these volumes and all chapter authors of these volumes have benefited from her theoretical framework.

Book Title: Re-defining Diversity and Dynamism of Natural Resource Management in Asia

Book Editors: Ganesh P. Shivakoti, Shubhechchha Sharma and Raza Ullah

Other volumes published:

1) Sustainable Natural Resources Management in Dynamic Asia Volume 1
   Ganesh P. Shivakoti, Ujjwal Pradhan and Helmi; editors
2) Upland Natural Resources and Social Ecological Systems in Northern Vietnam Volume 2
   Mai Van Thanh, Tran Duc Vien, Stephen J. Leisz and Ganesh P. Shivakoti; editors
3) Natural Resource Dynamics and Social Ecological System In Central Vietnam: Development, Resource Changes and Conservation Issues Volume 3

Tran N. Thang, Ngo T. Dung, David Hulse, Shubhechchha Sharma and Ganesh P. Shivakoti; editors

REDEFINING DIVERSITY AND DYNAMICS OF NATURAL RESOURCES MANAGEMENT IN ASIA

The Reciprocal Relationship between Governance of Natural Resources and Socio-Ecological Systems Dynamics in West Sumatra Indonesia

---

VOLUME 4

EDITED BY

RUDI FEBRIAMANSYAH, YONARIZA, RAZA ULLAH,
AND GANESH P. SHIVAKOTI

ELSEVIER

AMSTERDAM • BOSTON • HEIDELBERG • LONDON • NEW YORK • OXFORD
PARIS • SAN DIEGO • SAN FRANCISCO • SINGAPORE • SYDNEY • TOKYO

Elsevier
Radarweg 29, PO Box 211, 1000 AE Amsterdam, Netherlands
The Boulevard, Langford Lane, Kidlington, Oxford OX5 1GB, United Kingdom
50 Hampshire Street, 5th Floor, Cambridge, MA 02139, United States

Notices
Knowledge and best practice in this field are constantly changing. As new research and experience broaden our
understanding, changes in research methods, professional practices, or medical treatment may become necessary.

Practitioners and researchers must always rely on their own experience and knowledge in evaluating and using
any information, methods, compounds, or experiments described herein. In using such information or methods
they should be mindful of their own safety and the safety of others, including parties for whom they have a
professional responsibility.

To the fullest extent of the law, neither the Publisher nor the authors, contributors, or editors, assume any liability
for any injury and/or damage to persons or property as a matter of products liability, negligence or otherwise, or
from any use or operation of any methods, products, instructions, or ideas contained in the material herein.

Library of Congress Cataloging-in-Publication Data
A catalog record for this book is available from the Library of Congress

British Library Cataloguing-in-Publication Data
A catalogue record for this book is available from the British Library

ISBN: 978-0-12-805451-2

For information on all Elsevier publications
visit our website at https://www.elsevier.com/

Working together
to grow libraries in
developing countries

www.elsevier.com • www.bookaid.org

Publisher: Candice G. Janco
Acquisition Editor: Laura S Kelleher
Editorial Project Manager: Emily Thomson
Production Project Manager: Mohanapriyan Rajendran
Cover Designer: Matthew Limbert

Typeset by SPi Global, India

Contents

I

LIVELIHOOD DEPENDENCE, RIGHTS AND ACCESS TO NATURAL RESOURCES

1. Challenges of Managing Natural Resources in West Sumatra Indonesia

R. ULLAH, R. FEBRIAMANSYAH, YONARIZA

2. Methodological Approaches in Natural Resource Management

R. ULLAH

3. Livelihood Change and Livelihood Sustainability in the Uplands of Lembang Subwatershed, West Sumatra Province of Indonesia, in a Changing Natural Resources Management Context

MAHDI, G. SHIVAKOTI, D. SCHMIDT-VOGT

4. A Case Study of Livelihood Strategies of Fishermen in Nagari Sungai Pisang, West Sumatra, Indonesia

R. DESWANDI

II

TOWARDS EFFECTIVE MANAGEMENT OF CPRS

III

SOCIOECOLOGICAL SYSTEMS AND NEW FORMS OF GOVERNANCE

Contributors

R. Deswandi National Project Manager at the United Nations Industrial Development Organization, Jakarta, Indonesia

N. Effendi Andalas University, Padang, Indonesia

R. Febriamansyah Andalas University, Padang, Indonesia

Fetriyuna University of Padjadjaran, Bandung, Indonesia

D. Fiantis Andalas University, Padang, Indonesia

Helmi Andalas University, Padang, Indonesia

Hermansah Andalas University, Padang, Indonesia

S. Husin Andalas University, Padang, Indonesia

M.N. Janra Andalas University, Padang, Indonesia

S. Karimi Andalas University, Padang, Indonesia

Mahdi Andalas University, Padang, Indonesia

Nofriyanti Regional Body for Planning and Development, District of Padang Pariaman, Indonesia

S. Olivia Tito Andalas University, Padang, Indonesia

U. Pradhan World Agro-Forestry Center, Bogor, Indonesia

A. Saptomo University of Pancasila, Jakarta, Indonesia

S.M. Sari Freelance consultant, Bukittinggi, Indonesia

D. Schmidt-Vogt Asian Institute of Technology, Bangkok, Thailand

S. Sharma WWF-Nepal, Kathmandu, Nepal

G. Shivakoti Asian Institute of Technology, Bangkok, Thailand; The University of Tokyo, Tokyo, Japan

R. Ullah The University of Agriculture, Peshawar, Pakistan

E.L. Webb National University of Singapore, Singapore, Singapore

Yenni Andalas University, Padang, Indonesia

Yolamalinda STKIP PGRI, West Sumatra, Indonesia

Yonariza Andalas University, Padang, Indonesia

Yuerlita Andalas University, Padang, Indonesia

Words From Book Editors

CONTEXT

Elinor Ostrom received the Nobel Prize in Economics for showing how the "commons" is vital to the livelihoods of many throughout the world. Her work examined the rhetoric of the "tragedy of the commons," which has been used as the underlying foundation in privatizing property and centralizing its management as a way to protect finite resources from depletion. She worked, along with others, to overturn the "conventional wisdom" of the tragedy of the commons by validating the means and ways that local resources can be effectively managed through common property regimes instead of through the central government or privatization. Ostrom identified eight design principles relating to how common pool resources can be governed sustainably and equitably in a community. Similarly, the Institutional Analysis and Development (IAD) framework summarizes the ways in which institutions function and adjust over time. The framework is a "multilevel conceptual map," which describes a specific hierarchical section of interactions made in a system. The framework seeks to identify and explain interactions between actors and action situations.

As a political scientist, Ostrom has been a source of inspiration for many researchers and social scientists, including this four volumes book. Her theories and approach serve as the foundation for many of the chapters within these volumes. Following in her footsteps, the books is based on information collected during fieldwork that utilized quantitative as well as qualitative data, and on comparative case studies, which were then analyzed to gain an understanding of the situation, rather than starting from a formulated assumption of reality. The case studies in these volumes highlight the issues linked to the management of the environment and natural resources, and seek to bring about an understanding of the mechanisms used in managing the natural resource base in the regions, and how different stakeholders interact with each other in managing these natural resources. The details of the book are as follows:

Volume title		Editors
"Re-defining Diversity and Dynamism of Natural Resources Management in Asia"		Ganesh P. Shivakoti, Shubhechchha Sharma, and Raza Ullah
Volume I	Sustainable Natural Resources Management in Dynamic Asia	Ganesh P. Shivakoti, Ujjwal Pradhan, and Helmi
Volume II	Upland Natural Resources and Social Ecological Systems in Northern Vietnam	Mai Van Thanh, Tran Duc Vien, Stephen J. Leisz, and Ganesh P. Shivakoti
Volume III	Natural Resource Dynamics and Social Ecological Systems in Central Vietnam: Development, Resource Changes and Conservation Issue	Tran Nam Thang, Ngo Tri Dung, David Hulse, Shubhechchha Sharma, and Ganesh P. Shivakoti

Continued

Volume title		Editors
Volume IV	Reciprocal Relationship between Governance of Natural Resources and Socio-Ecological Systems Dynamics in West Sumatra Indonesia	Rudi Febriamansyah, Yonariza, Raza Ullah, and Ganesh P. Shivakoti

These volumes are made possible through the collaboration of diverse stakeholders. The intellectual support provided by Elinor Ostrom and other colleagues through the Ostrom Workshop in Political Theory and Policy Analysis at the Indiana University over the last two and half decades has provided a solid foundation for drafting the book. The colleagues at the Asian Institute of Technology (AIT) have been actively collaborating with the Workshop since the creation of the Nepal Irrigation, Institutions and Systems (NIIS) database; and the later Asian Irrigation, Institutions and Systems (AIIS) database (Ostrom, Benjamin and Shivakoti, 1992; Shivakoti and Ostrom, 2002; Shivakoti et al., 2005; Ostrom, Lam, Pradhan and Shivakoti, 2011). The International Forest Resources and Institutions (IFRI) network carried out research to support policy makers and practitioners in designing evidence based natural resource polices based on the IAD framework at Indiana University, which was further mainstreamed by the University of Michigan. In order to support this, the Ford Foundation (Vietnam, India, and Indonesia) provided grants for capacity building and concerted knowledge sharing mechanisms in integrated natural resources management (INRM) at Indonesia's Andalas University in West Sumatra, Vietnam's National University of Agriculture (VNUA) in Hanoi, and the Hue University of Agriculture and Forestry (HUAF) in Hue, as well as at the AIT for collaboration in curric-

ulum development and in building capacity through mutual learning in the form of masters and PhD fellowships (Webb and Shivakoti, 2008). Earlier, the MacArthur Foundation explored ways to support natural resource dependent communities through the long term monitoring of biodiversity, the domestication of valuable plant species, and by embarking on long-term training programs to aid communities in managing natural resources.

VOLUME 1

This volume raises issues related to the dependence of local communities on natural resources for their livelihood; their rights, access, and control over natural resources; the current practices being adopted in managing natural resources and socio-ecological systems; and new forms of natural resource governance, including the implementation methodology of REDD+ in three countries in Asia. This volume also links regional issues with those at the local level, and contributes to the process of application of various multimethod and modeling techniques and approaches, which is identified in the current volume in order to build problem solving mechanisms for the management of natural resources at the local level. Earlier, the Ford Foundation Delhi office supported a workshop on Asian Irrigation in Transition, and its subsequent publication (Shivakoti et al., 2005) was followed by Ford Foundation Jakarta office's long term support for expanding the knowledge on integrated natural resources management, as mediated by institutions in the dynamic social ecological systems.

VOLUME 2

From the early 1990s to the present, the Center for Agricultural Research and Ecological Studies (CARES) of VNUA and the School of Environment, Resources and Development (SERD) of AIT have collaborated in studying and understanding the participatory process that has occurred during the transition from traditional swidden farming to other farming systems promoted as ecologically sustainable, livelihood adaptations by local communities in the northern Vietnamese terrain, with a special note made to the newly emerging context of climate change. This collaborative effort, which is aimed at reconciling the standard concepts of development with conservation, has focused on the small microwatersheds within the larger Red River delta basin. Support for this effort has been provided by the Ford Foundation and the MacArthur Foundation, in close coordination with CARES and VNUA, with the guidance from the Ministry of Agriculture and Rural Development (MARD) and the Ministry of Natural Resources and Environment (MONRE) at the national, regional, and community level. Notable research documentation in this volume includes issues such as local-level land cover and land use transitions, conservation and development related agro-forestry policy outcomes at the local level, and alternative livelihood adaptation and management strategies in the context of climate change. A majority of these studies have examined the outcomes of conservation and development policies on rural communities, which have participated in their implementation through collaborative governance and participatory management in partnership with participatory community institutions. The editors and authors feel that the findings of these rich field-based studies will not only be of interest and use to national policymakers and practitioners and the faculty and students of academic institutions, but can also be equally applicable to guiding conservation and development issues for those scholars interested in understanding a developing

country's social ecological systems, and its context-specific adaptation strategies.

VOLUME 3

From the early 2000 to the present, Hue University of Agriculture and Forestry (HUAF) and the School of Environment, Resources and Development (SERD) of AIT supported by MacArthur Foundation and Ford Foundation Jakarta office have collaborated in studying and understanding the participatory process of Social Ecological Systems Dynamics that has occurred during the opening up of Central Highland for infrastructure development. This collaborative effort, which is aimed at reconciling the standard concepts of development with conservation, has focused on the balance between conservation and development in the buffer zone areas as mediated by public resource management institutions such as Ministry of Agriculture and Rural Development (MARD), Ministry of Natural Resources and Environment (MONRE) including National Parks located in the region. Notable research documentation in this volume includes on issues such as local level conservation and development related policy outcomes at the local level, alternative livelihood adaptation and management strategies in the context of climate change. A majority of these studies have examined the outcomes of conservation and development policies on the rural communities which have participated in their implementation through collaborative governance and participatory management in partnership with participatory community institutions.

VOLUME 4

The issues discussed above are pronounced more in Indonesia among the Asian countries and the Western Sumatra is such typical example mainly due to earlier logging concessionaries, recent expansion of State and private plantation of para-rubber and oil palm plantation. These new frontiers have created confrontations among the local community deriving their livelihoods based on inland and coastal natural resources and the outsiders starting mega projects based on local resources be it the plantations or the massive coastal aqua cultural development. To document these dynamic processes Ford Foundation Country Office in Jakarta funded collaborative project between Andalas University and Asian Institute of Technology (AIT) on Capacity building in Integrated Natural Resources Management. The main objective of the project was Andalas faculty participate in understanding theories and diverse policy arenas for understanding and managing common pool resources (CPRs) which have collective action problem and dilemma through masters and doctoral field research on a collaborative mode (AIT, Indiana University and Andalas). This laid foundation for joint graduate program in Integrated Natural Resources Management (INRM). Major activities of the Ford Foundation initiatives involved the faculty from Andalas not only complete their degrees at AIT but also participated in several collaborative training.

1 BACKGROUND

Throughout Asia, degradation of natural resources is happening at a higher rate, and is a primary environmental concern. Recent tragedies associated with climate change have left a clear footprint on them, from deforestation, land degradation, and changing hydrological and precipitation patterns. A significant proportion of land use conversion is undertaken through rural activities, where resource degradation and deforestation is often the result of overexploitation by users who make resource-use decisions based on a complex matrix of options, and potential outcomes.

South and Southeast Asia are among the most dynamic regions in the world. The fundamental political and socioeconomic setting has been altered following decades of political, financial, and economic turmoil in the region. The economic growth, infrastructure development, and industrialization are having concurrent impacts on natural resources in the form of resource degradation, and the result is often social turmoil at different scales. The natural resource base is being degraded at the cost of producing economic output. Some of these impacts have been offset by enhancing natural resource use efficiency, and through appropriate technology extension. However, the net end results are prominent in terms of increasing resource depletion and social unrest. Furthermore, climate change impacts call for further adaptation and mitigation measures in order to address the consequences of erratic precipitation and temperature fluctuations, salt intrusions, and sea level increases which ultimately affect the livelihood of natural resource dependent communities.

Governments, Non-governmental organizations (NGOs), and academics have been searching for appropriate policy recommendations that will mitigate the trend of natural resource degradation. By promoting effective policy and building the capacity of key stakeholders, it is envisioned that sustainable development can be promoted from both the top-down and bottom-up perspectives. Capacity building in the field of natural resource management, and poverty alleviation is, then, an urgent need; and several policy alternatives have been suggested (Inoue and Shivakoti, 2015; Inoue and Isozaki, 2003; Webb and Shivakoti, 2008).

The importance of informed policy guidance in sustainable governance and the management of common pool resources (CPRs), in general, have been recognized due to the conflicting and competing demand for use of these resources in the changing economic context in Asia (Balooni and Inoue, 2007; Nath, Inoue and Chakma, 2005; Pulhin, Inoue and Enters, 2007; Shivakoti and Ostrom, 2008; Viswanathan and Shivakoti, 2008). This is because these resources are unique in respect to their context. The management of these resources are by the public, often by local people, in a partnership between the state and the local community; but on a day-to-day basis, the benefits are at the individual and private level. In the larger environmental context, however, the benefits and costs have global implications. There are several modes of governance and management arrangement possible for these resources in a private-public partnership. Several issues related to governance and management need to be addressed, which can directly feed into the ongoing policy efforts of decentralization and poverty reduction measures in South and South East Asia.

While there has been a large number of studies, and many management prescriptions made, for the management of natural resources, either from the national development point-of-view or from the local-level community perspectives, there are few studies which point toward the interrelationship among other resources and CPRs, as mediated by institutional arrangement, and that have implications for the management of CPRs in an integrated manner, vis-a-vis poverty reduction. In our previous research, we have identified several anomalies and tried to explain these in terms of better management regimes for the CPRs of several Asian countries (Dorji, Webb and Shivakoti, 2006; Gautam, Shivakoti and Webb, 2004; Kitjewachakul, Shivakoti and Webb, 2004;

Mahdi, Shivakoti and Schmidt-Vogt, 2009; Shivakoti et al., 1997; Dung and Webb, 2008; Yonariza and Shivakoti, 2008). However, there are still several issues, such as the failure to comprehend and conceptualize social and ecological systems as coupled systems that adapt, self-organize, and are coevolutionary. The information obtained through these studies tends to be fragmented and scattered, leading to incomplete decision making, as they do not reflect the entire scenario. The shared vision of the diverse complexities, that are the reality of natural resource management, needs to be fed into the governance and management arrangements in order to create appropriate management guidelines for the integrated management of natural resources, and CPR as a whole.

Specifically, the following issues are of interest:

a. How can economic growth be encouraged while holding natural resources intact?
b. How has the decentralization of natural management rights affected the resource conditions, and how has it addressed concerns of the necessity to incorporate gender concerns and social inclusion in the process?
c. How can the sustainability efforts to improve the productive capacity of CPR systems be assessed in the context of the current debate on the effects of climate change, and the implementation of new programs such as Payment for Ecosystem Services (PES) and REDD+?
d. How can multiple methods of information gathering and analysis (eg use of various qualitative and quantitative social science methods in conjunction with methods from the biological sciences, and time series remote sensing data collection methods) on CPRs be integrated into national natural resource

policy guidelines, and the results be used by local managers and users of CPRs, government agencies, and scholars?

e. What are the effective polycentric policy approaches for governance and management of CPRs, which are environmentally sustainable and gender balanced?

2 OBJECTIVES OF THESE VOLUMES

At each level of society, there are stakeholders, both at the public and private level, who are primarily concerned with efforts of management enhancement and policy arrangements. Current theoretical research indicates that this is the case whether it is deforestation, resource degradation, the conservation of biodiversity hotspots, or climate change adaptation. The real struggles of these local-level actors directly affect the management of CPR, as well as the hundreds of people who are dependent upon them for a living. This book is about those decisions as the managers of natural resources. Basically, the authors of these chapters explore outcomes after decentralization and economic reforms, respectively. The volumes of this book scrutinize the variations of management practices with, and between, communities, local administration, and the CPR. Economic growth is every country's desire, but in the context of South and South East Asia, much of the economic growth is enabled by the over use of the natural resource base. The conundrum is that these countries need economic growth to advance, but the models of economic growth that are advanced, negatively affect the environment, which the country, depends upon. Examples of this are seen in such varied contexts as the construction of highways through protected areas, the construction of massive hydropower dams, and the conversion of traditional agricultural fields into rubber and oil palm plantations.

The research also shows that the different levels of communities, administration, and people are sometimes highly interactive and overlapping, for that reason, it is necessary to undertake coordinated activities that lead to information capture and capacity building at the national, district, and local levels. Thus the impacts of earlier intervention efforts (various policies in general and decentralization in particular) for effective outcomes have been limited, due to the unwillingness of higher administrative officials to give up their authority, the lack of trust and confidence of officials in the ability of local communities in managing CPR, local elites capturing the benefits of decentralization in their favor, and high occurrences of conflicts among multiple stakeholders at the local level (IGES, 2007).

In the areas of natural resource management particular to wildlife ecology monitoring and climate change adaptation, the merging of traditional knowledge with science is likely to result in better management results. Within many societies, daily practices and ways of life are constantly changing and adapting to new situations and realities. Information passed through these societies, while not precise and usually of a qualitative nature, is valued for the reason that it is derived from experience over time. Scientific studies can backstop local knowledge, and augment it through the application of rigorous scientific method derived knowledge, examining the best practices in various natural resource management systems over spatial and temporal scales. The amalgamation of scientific studies and local knowledge, which is trusted by locals, may lead to powerful new policies directed toward nature conservation and livelihood improvements.

Ethnic minorities, living in the vicinity to giant infrastructure projects, have unequal access, and control over, resources compared to other more powerful groups. Subsistence agriculture, fishery, swiddening, and a few off-farm options are the livelihood activities for these individuals. But unfortunately, these livelihood options are in areas that will be hit the most by changing climatic scenarios, and these people are the least equipped to cope; a situation that further aggravates the possibility of diversifying their livelihood options. Increasing tree coverage can help to mitigate climate change through the sequestering of carbon in trees. Sustainably planting trees requires technical, social, and political dimensions that are mainly possible through the decentralization of power to local communities to prevent issues of deforestation and degradation. The role of traditional institutions hence becomes crucial to reviving social learning, risk sharing, diversifying options, formulating adaptive plans and their effective implementation, fostering stress tolerance, and capacity building against climate change effects.

Though, the role of institutions in managing common pool resources has been explained in literature, it is also worth noting that institutions play significant roles in climate change adaptation. A study conducted by Gabunda and Barker (1995) and Nyangena (2004) observed that household affiliations in social networks were highly correlated with embracing soil erosion retaining technologies. Likewise, Jagger and Pender (2006) assumed that individuals involved in natural resource management focused programs were likely to implement land management expertise, regardless of their direct involvement in particular organizations. Friis-Hansen (2005) partially verifies that there is a positive relationship among participation in a farmer's institution and the adoption of smart agriculture technology. Dorward et al. (2009) correspondingly notes that institutions are vital in shaping the capability of local agrarians

to respond to challenges and opportunities. This study has also shown that institutions are the primary attribute in fostering individuals and households to diversify livelihoods in order to adapt to a changing climate. In the context of REDD+, a system is required that can transcend national boundaries, interconnect different governance levels, and allow both traditional and modern policy actors to cooperate. Such a system emphasizes the integration of both formal and informal rule making mechanisms and actor linkages in every governance stage, which steer toward adapting to and mitigating the effects of local and global environmental change (Corbera and Schroeder, 2010).

Based on the above noted discussions, the volumes in this book bring these issues forward for a global audience and policy makers. Though earlier studies show that the relationship between scientific study and outcomes in decision making are usually complex; we hope that the studies examined and discussed here can have some degree of impact on academics, practitioners, and managers.

G. Shivakoti, S. Sharma, and R. Ullah

References

Balooni, K.B., Inoue, M., 2007. Decentralized forest management in South and Southeast Asia. J. Forest. 2007, 414–420.

Corbera, E., Schroeder, H., 2010. Governing and Implementing REDD+. Environ. Sci. Pol. http://dx.doi.org/10.1016/j.envsci.2010.11.002.

Dorji, L., Webb, E., Shivakoti, G.P., 2006. Forest property rights under nationalized forest management in Bhutan. Environ. Conservat. 33 (2), 141–147.

Dorward, A., Kirsten, J., Omamo, S., Poulton, C., Vink, N., 2009. Institutions and the agricultural development challenge in Africa. In: Kirsten, J.F., Dorward, A.R., Poulton, C., Vink, N. (Eds.), Institutional Economics Perspectives on African Agricultural Development. IFPRI, Washington DC.

Dung, N.T., Webb, E., 2008. Incentives of the forest land allocation process: Implications for forest management in Nam Dong District, Central Vietnam. In: Webb, E., Shivakoti, G.P. (Eds.), Decentralization, Forests and Rural Communities: Policy outcome in South and South East Asia. SAGE Publications, New Delhi, pp. 269–291.

Friis-Hansen, E., 2005. Agricultural development among poor farmers in Soroti district, Uganda: Impact Assessment of agricultural technology, farmer empowerment and changes in opportunity structures. Paper presented at Impact Assessment Workshop at CYMMYT, Mexico, 19–21. October. http://citeseerx.ist.psu.edu/viewdoc/download?doi=10.1.1.464.8651&rep=rep1&type=pdf.

Gautam, A., Shivakoti, G.P., Webb, E.L., 2004. A review of forest policies, institutions, and the resource condition in Nepal. Int. Forest. Rev. 6 (2), 136–148.

Gabunda, F., Barker, R., 1995. Adoption of hedgerow technology in Matalom, Leyte Philipines. Mimeo. In: Bluffstone, R., Khlin, G. (Eds.), 2011. Agricultural Investment and Productivity: Building Sustainability in East Africa. RFF Press, Washington, DC/London.

IGES, 2007. Decentralization and State-sponsored Community Forestry in Asia. Institute for Global Environmental Studies, Kanagawa.

Inoue, M., Isozaki, H., 2003. People and Forest-policy and Local Reality in Southeast Asia, the Russian Far East and Japan. Kluwer Academic Publishers, Netherlands.

Inoue, M., Shivakoti, G.P. (Eds.), 2015. Multi-level Forest Governance in Asia: Concepts, Challenges and the Way Forward. Sage Publications, New Delhi/California/London/Singapore.

Jagger, P., Pender, J., 2006. Impacts of Programs and Organizations on the Adoption of Sustainable Land Management Technologies in Uganda. IFPRI, Washington, DC.

Kijtewachakul, N., Shivakoti, G.P., Webb, E., 2004. Forest health, collective behaviors and management. Environ. Manage. 33 (5), 620–636.

Mahdi, Shivakoti, G.P., Schmidt-Vogt, D., 2009. Livelihood change and livelihood sustainability in the uplands of Lembang Subwatershed, West Sumatra, Indonesia, in a changing natural resource management context. Environ. Manage. 43, 84–99.

Nath, T.K., Inoue, M., Chakma, S., 2005. Prevailing shifting cultivation in the Chittagong Hill Tracts, Bangladesh: some thoughts on rural livelihood and policy issues. Int. For. Rev. 7 (5), 327–328.

Nyangena, W., 2004. The effect of social capital on technology adoption: empirical evidence from Kenya. Paper presented at 13th Annual Conference of the European Association of Environmental and Resource Economics, Budapest.

Ostrom, E., Benjamin, P., Shivakoti, G.P., 1992. Institutions, Incentives, and Irrigation in Nepal: June 1992. (Monograph) Workshop in Political Theory and Policy Analysis, Indiana University, Bloomington, Indiana, USA.

Ostrom, E., Lam, W.F., Pradhan, P., Shivakoti, G.P., 2011. Improving Irrigation Performance in Asia: Innovative Intervention in Nepal. Edward Elgar Publishers, Cheltenham, UK.

Pulhin, J.M., Inoue, M., Enters, T., 2007. Three decades of community-based forest management in the Philippines: emerging lessons for sustainable and equitable forest management. Int. For. Rev. 9 (4), 865–883.

Shivakoti, G., Ostrom, E., 2008. Facilitating decentralized policies for sustainable governance and management of forest resources in Asia. In: Webb, E., Shivakoti, G.P. (Eds.), Decentralization, Forests and Rural Communities: Policy Outcomes in South and Southeast Asia. Sage Publications, New Delhi/Thousand Oaks/London/Singapore, pp. 292–310.

Shivakoti, G.P., Ostrom, E. (Eds.), 2002. Improving Irrigation Governance and Management in Nepal. Institute of Contemporary Studies (ICS) Press, California, Oakland.

Shivakoti, G.P., Vermillion, D., Lam, W.F., Ostrom, E., Pradhan, U., Yoder, R., 2005. Asian Irrigation in Transition-Responding to Challenges. Sage Publications, New Delhi/Thousand Oaks/London.

Shivakoti, G., Varughese, G., Ostrom, E., Shukla, A., Thapa, G., 1997. People and participation in sustainable development: understanding the dynamics of natural resource system. In: Proceedings of an International Conference held at Institute of Agriculture and Animal Science, Rampur, Chitwan, Nepal. 17–21 March, 1996. Bloomington, Indiana and Rampur, Chitwan.

Viswanathan, P.K., Shivakoti, G.P., 2008. Adoption of rubber integrated farm livelihood systems: contrasting empirical evidences from Indian context. J. For. Res. 13 (1), 1–14.

Webb, E., Shivakoti, G.P. (Eds.), 2008. Decentralization, Forests and Rural Communities: Policy Outcomes in South and Southeast Asia. Sage Publications, New Delhi/Thousand Oaks/London/Singapore.

Yonariza, Shivakoti, G.P., 2008. Decentralization and co-management of protected areas in Indonesia. J. Legal Plur. 57, 141–165.

Foreword

The shared concern for rapidly depleting and degenerating natural resources is a worldwide phenomenon that has brought researchers and politicians together in uneasy cooperation. For a long time, research on natural resources was compartmentalized according to distinct resources. Now, perspectives are slowly shifting towards a more integrated approach. Researchers have pointed to the necessity of taking the totality of natural resources into consideration in tandem with diverse uses. They have called attention to the fact that natural resources are always situated within specific, but dynamic, socio-ecological contexts that call for locally suitable solutions. With her decades-long collaborative work on integrated water management and common property regimes and due in no small part to her keen eye for the political implications, Elinor Ostrom has been at the forefront of the movement to stimulate such integrated research.

This is the fourth volume in a series testifying to the immense inspiration that Elinor Ostrom instilled into so many researchers devoted to the study of natural resources in Southeast Asia. Focusing on one single region in Indonesia, West Sumatra, the volume covers studies of a wide range of natural resources, including fisheries, diverse forms of water use (drinking water, health and sanitation, irrigation, hydroelectricity), and forests. It is a result of the long-term involvement of a lively group of researchers from Andalas University in collaboration with Elinor Ostrom and a number of other research institutes and donor agencies interested in natural resource management. This volume contains contributions not just of highly experienced scholars, but also of young researchers from Andalas University. Together, their contributions demonstrate the virtues of long-term collaborations for developing a suitable framework with a broad perspective on natural resource management.

The focus on West Sumatra is well chosen for a number of reasons. It is a region with substantial forest resources that have come under severe pressure, due to a long history of logging concessions, the recent expansion of plantations, rapidly increasing population, and economic hardship that has led to increasing levels of illegal logging, resulting in disappearing forest areas and deteriorating land and water resources. The volume provides comprehensive insights into the complexities of managing forest areas, water resources, and fisheries, and demonstrates how intimately these resources are related. The contributions of the researchers build upon the broad political–economic perspective that Elinor Ostrom and her group has developed over the past decades.

A second reason why West Sumatra is exceptionally interesting is that it was among the first regions in which Indonesia's decentralization policies were put into practice at district and lower levels. Since the 1990s, decentralization policies have been proliferating throughout the world. Initially, the idea behind it was that a decentralized government would be more democratic, participative, and accountable. If people would be more involved in the management of their natural resources, they would be more prudent in their use. Importantly, decentralization would also significantly help to combat rampant corruption. But critics have argued that even corruption would be decentralized and decentralization would put the onus of protection and conservation on local populations, without endowing them with the capacity to fully profit from their natural resources. West Sumatra is an important example to test these propositions because it actively captures the process rather than imposing a top-down policy. In the process of decentralization, the region decided to revert to a village-administrative structure, which incorporated salient features of *adat*, the customary law of the region, similar to the structure that had been operative until 1983, when it was abandoned and a unified administrative structure was imposed throughout the country. As a result, villages in West Sumatra actively engaged in implementing the decentralization process and village governments took an exceptional interest in the governance of natural resources within their territory. As this volume shows, the recent history of West Sumatra offers unique opportunities to explore the possibilities and constraints of natural resource management under the condition of the active participation of local communities alongside other levels of governance. The various contributions demonstrate that the process of decentralization has generated highly diverse responses that suggest that solutions to natural resource problems must be tailored to the demands of specific situations—for uniform solutions are bound to fail. All the chapters take a positive view of active local participation, however, without idealizing local communities. The volume insists that neither local communities nor the state can possibly solve the immense problems without the cooperation of the other and that multilevel policies are needed. The chapters also call for strengthening local institutions and for increasing awareness and knowledge among the general population in this regard.

The significance of many questions raised in this volume extend beyond the specificities of West Sumatra and deserve to be explored further in other contexts. At the same time, some questions raised here give rise to others and can generate new perspectives and research trajectories. Various chapters mention that corruption and illegal logging have increased since the introduction of decentralization. The question is to what extent this is a result of decentralization, as is often argued. Could there be other factors, such as the decades-long tight-grip of the armed forces that have stood in the way of the development of strong civil institutions? The volume suggests that policies concerning natural resource management affect different sections of the population differently. There seems to be a dividing line between generations, and that deserves more research. Besides, the chapter on gender suggests that a great deal needs to be done with regard to gender mainstreaming, both for ensuring proper access to and for the management of natural resources. What exactly would be the difference if gender issues were taken seriously? What are the crucial constraints that prevent such mainstreaming from taking place? Several chapters also point to the danger of local elites and outsiders capturing the natural resources. This raises the question of how and why this might happen. How serious is this danger? Is it indeed due to decentralization, or are there other factors in play? And what might be done to prevent or mitigate elite and outsider capture? Moreover, the call to strengthen local institutions begs the question of what this actually means. What kinds of institutions are needed, what should be their authority and their competence, and in what way do they need strengthening? How might that be achieved and what are the dynamics and constraints of such processes? What exactly can be learned from the positive examples that are presented in this volume? Finally, the chapters show how the plural legal order, which includes customary law (*adat*), and state regulations at all levels, from the national down to the village level, form the complex legal landscape in which users and managers of natural resources have to navigate. Several chapters point to the importance of clear and secure land titles, though the volume suggests that this might not be the most important problem. The chapters demonstrate that poorly designed, overlapping, and contradictory legislation constitutes a formidable constraint to the management of natural resources, not to speak of the seriously deficient implementation. The chapters argue that this has a deep impact on local natural resource use and management. In particular, corruption and deficient implementation have led to considerable disappointment, and that in turn has decreased the willingness among local people to cooperate. Further local research might provide a deeper insight into the concrete local impact of deficient legislation and the plural legal order. However, the kind of research and policy making that is necessary to redress these deficiencies would primarily concern higher levels of governance and less local regulations. An important question for further research relates to processes in which policy makers and researchers collaborate, in order to better understand the uneasy relationships, the misunderstandings, as well as the different interests and aims that make collaboration so difficult.

These are all difficult, yet extremely important, questions. The great strength of this volume is that it offers clear state-of-the-art perspectives and raises many pressing issues, while offering important suggestions as to the kinds of improvements that are needed. Future research and policymaking can build on this foundation. Redressing the problems of integrated natural resource management will remain an ongoing and dynamic process, in which, inevitably, new questions will have to be addressed and adjustments made to the ever-changing political, legal, social, economic, and ecological constellations. The volume brings to light a dynamic approach to integrated resource management.

Keebet von Benda-Beckmann
Max Planck Institute for Social
Anthropology, Halle/Saale, Germany

Preface

Eco- and social-systems are two interrelated components in natural resources governance and management. The key challenge of good governance and management is how to balance between the two in order to maintain the flow of benefit-streams from natural resources. In such situations, understanding the reciprocal relationship between governance and the socio-ecological system is very important.

The need to create an effective governance system requires a deeper understanding of various factors involved in the interface between the eco- and social-systems. This book identified various factors from research conducted in West Sumatra, Indonesia. West Sumatra is a unique region to study the various aspects of effective natural resources governance for a number of reasons. First, it has different types of natural resources (land, water, forest, and coastal/sea) in a relatively smaller area, which provides an opportunity to understand complete pictures of natural resources management. Second, historically it has strong socio-cultural and religious practices related to natural resources management, which provide opportunities to understand how the socio-cultural and religious values are translated into practices. Third, West Sumatra, together with other provinces in Indonesia, has implemented the decentralization policy of the government, which provides opportunities to understand the interface between national policy and laws in order to recognize and integrate the existing socio-cultural and religious value. Fourth, the majority of the population is dependent on land, water, forest, and coastal/sea

resource-based livelihood activities, which give an opportunity to understand how the socio-economic interests were reconciled with the need to conserve natural resources.

Some of the important issues commonly raised by most of the authors are the roles of local institutions (developed and based on socio-cultural and religious values), the need to give attention to the socio-economic aspects of the household (livelihood), awareness and knowledge of the community on the relationship between their livelihoods and care of the environment, tenurial aspects, and policies on co-management. All of these findings place emphasis on the need for integrated, participatory, socio-cultural, and religious-based approaches in natural resources management, both in the aspects and level of governance. In addition, all the cases analyzed in this volume have recognized and addressed the cross-scale dynamics in designing and implementing an effective natural resource governance arrangement by examining the human-nature relationship through combination of approaches, methods, and techniques. The methodologies used to assess the multilevel governance of natural resources, cross-scale coordination, and the application of multimethods in natural resource management have the imprint of Ostrom's framework and analytic approaches.

This volume has been made possible through the direct and indirect contributions of several organizations and individuals. We wish to acknowledge the support provided by the Ford Foundation country office in Jakarta, which paved the way for academic collaboration between the Asian Institute of Technology and Andalas University, with the intellectual support provided by The Ostrom Workshop in Political Theory and Policy Analysis of Indiana University. We wish to acknowledge tremendous efforts of the colleagues from Ostrom Center for Advanced Study in Natural Resources Governance (OCeAN) of Asian Institute of Technology who shouldered the burden of bringing this volume to print. We appreciate the endless administrative and editing efforts of Laura Kelleher and Emily Thomson of Elsevier to put everything into a coherent order.

Helmi
Researcher at the Sustainable Development and Sustainability Science Initiative, Andalas University, Padang, Indonesia

Ujjwal Pradhan
World Agro-Forestry Center, SE Asian Regional Office, Bogor, Indonesia

LIVELIHOOD DEPENDENCE, RIGHTS AND ACCESS TO NATURAL RESOURCES

Challenges of Managing Natural Resources in West Sumatra Indonesia

R. Ullah*, R. Febriamansyah†, Yonariza†
*The University of Agriculture, Peshawar, Pakistan
†Andalas University, Padang, Indonesia

1.1 INTRODUCTION

1.1.1 Background

Sumatra Island with its seven provinces in Indonesia is one of the largest islands of the country and is home to approximately one-quarter of the country's population. Sumatra Island is also a global hotspot for biodiversity and is home to the world's last remaining and endangered species including Sumatran tigers, elephants, and rhinos. The natural environment of the island provides fertile land for agriculture and forests that sustain native species and local communities. The availability of plentiful natural resources support the production of palm oil, paper, and coffee and also provide clean water and confiscate a large amount of carbon in forests and peatlands. In Indonesia, West Sumatra Province along with Papua are the most advanced provinces in the nation's decentralization process (Kurauchi et al., 2006). In West Sumatra more than half of the land is declared as forest area due to its topographic conditions; that is, mountainous areas where most of the forest belongs to conservation forest and protection forest. The province is the land of Minangkabau, a contemporary, largely matrilineal society where social organization has taken a complete shape called *nagari* and each has its own territory with clear boundary. Most of the areas declared by central government as forest areas overlap with *nagari* territory. As an autonomous social unit it governs resources within its territory, but there is also wide variation of local forest management policy in each *nagari* depending on the type of forest found in each *nagari*. Some *nagaris* are dominated by conservation forest, while others are dominated by protection forest, and a few *nagaris* fall under the production forest category. The presence of these forests play a critical role nationally by providing ecosystem services for rural communities and internationally by maintaining biodiversity and contributing to the regulation of climatic systems (McKay, 2013). However, Indonesia in general and Sumatra Island in particular, have high rates of tropical deforestation, especially beyond protected areas (Measey, 2010). The high rate of deforestation is causing forest habitat to become increasingly fragmented and isolated, thereby increasing the vulnerability of endangered species, particularly tigers, to extinction (Linkie et al., 2007; Dinerstein et al., 2007). Thus, innovative approaches to complement mainstay community-based conservation programs are urgently needed (McKay, 2013).

Serious environmental challenges put huge pressure on the natural resource base of Sumatra and poses serious threats for the sustainability of these natural resources. The island has lost nearly half of its forest cover in recent decades, causing a major reduction in carbon stock and other sources of Indonesia's natural capital. Economic development has brought many benefits to the people of Sumatra; however, it has been accompanied by degraded water quality, massive greenhouse gas emissions, and increased erosion. Logging activities, deforestation, and land conversion to industrial plantations have depleted resources critical to wildlife and human well being.

1.1.2 Decentralization and Natural Resource Management in West Sumatra

In Indonesia the social and political situation has changed rapidly since the collapse of the centralized and authoritarian Suharto regime due to an economic crisis that hit Indonesia in the mid-1990s. The fall of the authoritarian New Order regime of Suharto in 1998 was followed by the advent of democratic decentralization in Indonesia. In 1999, the national government enacted Law No. 22/1999 on Regional Autonomy and Law No. 25/1999 on Fiscal Decentralization as part of a broader effort to curb the prevailing corruption, collusion, and nepotism of the Suharto era. Consequently, powers and responsibilities over a large number of government functions were devolved to district and municipal authorities. These included the sectors of health, education, public works, agriculture, communications, industry and trade, labor, capital investment, environment, and land affairs (Silver, 2003). In line with the central government's decision to uphold and strengthen local autonomy, governments at the subnational level have undertaken policy and administrative reforms in the natural resources sector. The government of Indonesia (GoI) issued and implemented Law No. 22/1999 regarding the decentralization of regional governments. In addition, Forestry Law No. 41/1999 and Water Resources Law No. 7/2004 replaced older laws. These new laws recognize the role of traditional rules and regulations in natural resources management. Decentralization of natural resources management takes place in two ways. First is to devolve property rights over natural resources to local communities. Second is to hand over the formal powers of government to its own subunits. Both ways of decentralization claim that outcomes will be more efficient, flexible, equitable, accountable, and participatory (Andersson et al., 2004).

West Sumatra Provincial Government Regulation No. 9/2000 was formulated as a supplement to the 1999 decentralization laws to reintroduce *nagari* and to place decision-making authority in their hands. *Nagari* had suffered a severe setback for more than two decades under the uniform village administration system during the New Order regime, which did not recognize customary laws, boundaries, and institutions, and eroded the unity of *nagari* populations (Thorburn, 2000). Since the issuance of the regional regulation, thousands of artificially created villages have been remerged into hundreds of *nagari*. This "back to the *nagari*" policy is hailed by West Sumatrans as a crucial step toward the resurgence of customary norms and the restoration of property rights over communal lands (McCarthy, 2002; Bebbington et al., 2004).

New decentralization laws and policies at the provincial and district levels have given discretionary powers to *nagari* governments. However, despite power residing at a local level, downward accountability is lacking, as decisions issued by the local government and *nagari* representatives were often not in line with the needs and aspirations of the communities. The interpretation of provincial laws at lower levels of government is also a challenge. In some cases, district-level ordinances have been passed, but *nagaris* have protested the policies because of a lack of participation in their formulation. Thus, even where decision-making power has been passed from the provincial level to more local government bodies, communities themselves do not appear to have a strong voice in decision making.

1.2 NATURAL RESOURCE ISSUES IN WEST SUMATRA

The resource-wise local management problems and issues related to forests, land, water resources, protected areas, and biodiversity in West Sumatra are summarized below.

1.2.1 Forests

The implementation of regional autonomy and authority devolution to district and regional governments, which was expected to empower local people in decision making and policy formulation for natural resources management, failed to produce significant benefits. It even increased deforestation and forest degradation through the number of forestry industrial permits granted without considering their socioecological consequences.

The annual rate of deforestation and forest degradation in Indonesia, increased from 700,000 ha in the mid-1980s to approximately 2.4 million at the beginning of 1999. Major causes of deforestation in East Kalimantan, one of the most forest-rich provinces in Indonesia, is overexploitation, particularly illegal logging and forest conversion for other utilizations, especially extensive oil palm and coal mining industries. Owing to large-scale commercial logging and agroindustrial development, vast areas of Sumatra, particularly in the lowland forest in South and Central Sumatra, have also been deforested during recent decades. Forest conversions for agricultural activities both for food crops and plantation crops have adverse impacts on natural resources as well as hydrological functions including landslides, loss of fertility due to erosion (on-site impacts), and sedimentation and water shortage in dry season (off-site impacts). The water shortage ultimately leads to conflicts among water users.

The implementation of regional autonomy that was expected to empower the local community in decision making and policy formulation and help to create product sustainability has generally not yielded promising results. In fact, forest and land governance by the local government under the regional autonomy is far from good governance principles such as no transparency, less participation, low accountability, and weak coordination. Moreover, local governments were unable to optimize their authority for increasing community welfare; their policies had even degraded the remaining forests. Financial policies for stimulating regional incomes to support local economic development have also aggravated forest and land governance. In such a situation a good approach is to implement collaborative forest governance (CFG) where government and nongovernment institutions (including local communities) are believed to be a key for successful resource management.

Despite the importance of forest in terms of sustaining water supplies, protecting the soils of important watersheds, minimizing the effects of catastrophic floods and landslides, and providing economic benefits including employment and income for population living around it, the local communities in Sangir, West Sumatra, negatively affect the sustainability of the forests in three different ways: (1) cultivating land inside the forests, (2) mining activities, and (3) harvesting forest products, particularly extraction of timber. Major causes of deforestation in Indonesia include land-use change and illegal logging.

1.2.2 Land Resources

Most of the local people in Indonesia depend on natural resources, particularly forest resources and agriculture, for their livelihoods. Land resources are under immense pressure as a result of greatly increased population, economic growth and physical development, deforestation, uncontrolled soil erosion, and the impact of opening land and shifting cultivation. This increasing pressure on land resources leads to deterioration of land resources and an increase in conflicts and loss of capacity of resources to maintain their functions. In West Sumatra, forests and land have long been at the center of conflicts between concessionaires and locals over the management, use, and protection of natural resources.

With increasing population the land demand for settlement and public infrastructure also increases. For this purpose, the conversion of agricultural land into settlement areas is on the rise, which will reduce rice paddy cultivation, a staple food for a majority of Indonesian citizens, and ultimately result in food insecurity. Moreover, the land-use change in the upland areas of watersheds create problems of water quantity, quality, and reliability downstream. These issues lead to land deterioration and increased pressure on the land resources and the urge for proper and sustainable management of land resources.

1.2.3 Water Resources

Water resources and related ecosystems are under threat from population growth, pollution, unsustainable use, land-use changes, climate change, and many other forces. The performance of the water supply and sanitation in Indonesia, despite government efforts to provide clean drinking water and sanitation, is comparatively low, causing widespread waterborne diseases and diarrhea.

River basin management and water allocation have increasingly become issues in West Sumatra as competition for water use between irrigated agriculture and other sectors of the economy increases. Integrated water resource management (IWRM) is an important development agenda to address institutional problems and capacity building for the use, control, preservation, and sustainability of water systems. The GoI is in the process of reforming its water resources management policy, putting IWRM principles into action (Helmi, 2001).

The increasing levels of deforestation along with population growth threaten water quality for existing development, and water availability for future development in West Sumatra. The condition of food self-sufficient status has become worse, which was not just caused by global climate changes but also by suffering a lack of capability to manage water, and a lack of sustainable financing management and deforestation in the upper watershed. The increase in nonagriculture demand, the competition for water uses will increase conflicts between the different types of users and increase water quality problems during low flow. To support future development in water resources and to promote long-term water resources conservation and preservation (Anshori, 2005), the need to improve current performance of water resources management is widely appreciated. In managing scarce water resources, a change in attitude and approach is seen as essential. Participatory learning and action methods surfaced as a distinct need for coordination at the river basin level.

1.2.4 Protected Areas and Biodiversity

The Tropical Rainforest Heritage of Sumatra site comprises three national parks and holds the greatest potential for long-term conservation of the distinctive and diverse biota of Sumatra, including many endangered species. The area is home to around 10,000 plant species, more than 200 mammal species, and some 580 bird species. Of the mammal species, 22 are Asian, not found elsewhere in the archipelago, and 15 are confined to the Indonesian region, including the endemic Sumatran orangutan. The site also provides biogeographic evidence of the evolution of the island (Brun et al., 2015).

Increasing demands for timber and agricultural products along with weak enforcement of laws have resulted in illegal logging and agricultural encroachment within Indonesia's protected areas (Levang et al., 2012). Extensive deforestation coupled with agricultural encroachment and road development plans in Indonesia contributes substantially to land-based global carbon emissions (Harris et al., 2012) and potentially high rates of biodiversity loss (Wilcove et al., 2013). The fundamental threatening processes are directly linked to the access provided by roads and the failure to effectively enforce existing laws. Road access facilitates illegal logging, encroachment, and poaching, all of which pose significant threats to the integrity of the component parks of the property. The recent decentralization trend in Indonesia presents new challenges on making decentralization work for conservation processes including biodiversity and protected areas management. Local responses to decentralization vary widely both across spatial and infrastructural dimensions as well as at macro, meso, and micro levels, which have affected local initiatives differently with respect to protected areas of comanagement. Previous studies on protected areas management in Indonesia after decentralization policies have produced very controversial results. It is broadly argued that decentralization triggers local government efforts to exploit the remaining forest resources to earn short-term revenues through timber cutting and/or converting forest areas into agricultural lands. Converting protected forest into production forest, to increase regional income from logging permits, is also on the rise. Timber harvesting from protected areas is a threat to tropical forests, causes environmental damages, and promotes corruption.

1.3 BRIEF OUTLINE AND SUMMARY OF ISSUES ADDRESSED IN THE VOLUME

This volume comprises 16 chapters dealing with diverse issues related to natural resources and their management in West Sumatra, Indonesia. The chapters are further grouped into three sections: (I) Livelihood dependence, rights, and access to natural resources; (II) Natural resources management practices; and (III) Socioecological systems and new forms of governance. Details of these sections are discussed below.

I. LIVELIHOOD DEPENDENCE, RIGHTS AND ACCESS TO NATURAL RESOURCES

Section I: Livelihood Dependence, Rights, and Access to Natural Resources

The sustainability of natural resources will be significantly affected by the sustainability of livelihoods of the community in the surrounding natural resources system. Human beings have to retain adequate knowledge of adaptive capacity to secure their livelihoods in a dynamic environment. Their capability to develop livelihood strategies depends on the livelihood assets they own or might be accessed by the available institutions (eg, customary law). The issues related to the rural community's dependence on natural resources for livelihoods are discussed in Chapters 3 and 4. Issues related to the community's rights to natural resources are discussed in Chapter 5, while Chapters 6 and 7 address issues related to gender inequality in a fishing family in a small-scale fishery and rural water supply project, respectively.

Section II: Toward Effective Management of Community Property Rights

Restoration of the *nagari* is meant to reestablish and formalize customary rules on natural resources management, and to encourage wider participation. The decentralization supports legal recognition of community-based property rights (CBPRs) and various types of community-based natural resource management (CBNRM) initiatives. New laws (Law No. 7/2004 on water resources, replacing Law No. 11/1974 and Law No. 41/1999 to replace the old one) provide scope for the local administration as well as for people at the local level to play a substantial role in forest and water management. The changed context of natural resources management has influenced household's access to capital assets including access to forest resources, forestland, and irrigation infrastructure and their participation in the institution of water management. This section highlight issues related to land-use change (Chapter 8), sustainable land-use practices (Chapter 9), and forest management and issues related to illegal logging (Chapter 10). Indonesia has the world's largest Muslim population (87% of its 240 million population) and religion has a strong influence on daily life (Mangunjaya and Mckay, 2012).

Section III: Socioecological Systems and New Forms of Governance

The Social-Ecological System (SES) framework, developed by Elinor Ostrom and her colleagues (Ostrom, 2007, 2009), incorporates large, decomposable sets of social and ecological attributes that potentially affect choices and outcomes in SESs. The SES framework is very useful for understanding and diagnosing problems in complex socioecological systems (see, eg, Ostrom and Cox, 2010) where outcomes arise as a result of the interplay of multiple attributes such as the combination of leadership, social capital, and spatial management plans (Gutierrez et al., 2011). The long-term goal of this framework is to facilitate the accumulation of knowledge and to build theory across diverse cases (Ostrom, 2007, 2009). The framework enables researchers from different disciplinary backgrounds to share a common vocabulary for the construction and testing of alternative theories and models (McGinnis and Ostrom, 2014). This section raises issues related to the existing socioecological systems and new forms of natural resources governance in West Sumatra. The degradation of natural resources and environmental problems are major concerns for the GoI as well as for the local government of Sumatra. In response to these environmental issues the government has initiated several rehabilitation and conservation activities. The GoI has developed local regulations on the implementation of an autonomy system after the change of governmental systems from centralization to decentralization. The *nagari* institutions have been reintroduced in West Sumatra through formulation of provincial regulations after decentralization and authority devolution to the local level.

The government has finalized guidelines toward achieving the emissions commitment through reforestation, plantation under community forest, and the development of *Hutan Tanaman Rakyat* (HTR). To propose justifiable community-based sustainable plantation forest management, HTR was familiarized by the state and REDD+ involvement with HTR is likely to achieve the anticipated outcomes. REDD+ is expected to benefit forest and local people; however, accomplishment of REDD+ is unlikely without involving natives in the implementation process. Moreover, the government is planning to implement REDD+ without acknowledging the fact that forest management is a socioecological interaction involving institutions, political pressures, and user's actions. Uncertainty over poorly defined forest tenure and incentives agreed upon by the consensus building process are other factors that need to be addressed.

1.4 CONCLUSION

West Sumatra Indonesia is blessed with plenty of natural resources including forest, land resources, water resources, protect areas and biodiversity. However, the haphazard harvest put huge pressure on the natural resource base and challenge the viability and sustainability of these natural resources in West Sumatra. This volume comprised of case studies identifying the key challenges and threats to natural resources management in West Sumatra Indonesia particularly after decentralization. Shared experiences and lessons learned from these case studies may serve as a basis for policy makers and practitioners to recognize the potential of West Sumatra's natural resources for ecological, social, and economic development; food security; poverty alleviation; and natural resources sustainability.

References

Andersson, K.P., Gibson, C.C., Lehoucq, F., 2004. The Politics of Decentralized Natural Resource Governance [Electronic Version]. PS Online, pp. 421–426. Retrieved from www.apsanet.org.

Anshori, I., 2005. Basin Water Resources Management and Organization in Indonesia. NARBO Training Workshop in River Basin Management and Organizations for Mid-career Water Professionals from South Asia, Colombo.

Bebbington, A., Dharmawan, L., Fahmi, E., Guggenheim, S., 2004. Village politics, culture and community-driven development: insights from Indonesia. Progr. Develop. Stud. 4 (3), 187–205.

Brun, C., Cook, A.R., Lee, J.S.H., Wich, S.A., Koh, L.P., Carrasco, L.R., 2015. Analysis of deforestation and protected area effectiveness in Indonesia: A comparison of Bayesian spatial models. Glob. Environ. Chang. 31, 285. http://dx.doi.org/10.1016/j.gloenvcha.2015.02.004.

Dinerstein, E., Loucks, C., Wikramanayake, E., et al., 2007. The fate of wild tigers. Bioscience 57, 508–514.

Gutierrez, N.L., Hilborn, R., Defeo, O., 2011. Leadership, social capital and incentives promote successful fisheries. Nature 470 (7334), 386–389.

Harris, N.L., Brown, S., Hagen, S.C., Saatchi, S.S., Petrova, S., Salas, W., Hansen, M.C., Potapov, P.V., Lotsch, A., 2012. Baseline map of carbon emissions from deforestation in tropical regions. Science 336, 1573–1576.

Helmi, 2001. Strategies for improving the productivity of agricultural water management. In: Proceedings of the Regional Workshop, Malang, Indonesia, January 15–19, 2001.

Kurauchi, Y., La Vina, A., Badenoch, N., Fransen, L., 2006. Decentralization of Natural Resources Management: Lessons from Southeast Asia—Case Studies under REPSI. World Resource Institute, (http://www.wri.org/sites/default/files/pdf/repsisynthesis.pdf. Accessed 22nd August, 2014).

Levang, P., Sitorus, S., Gaveau, D.L.A., Sunderland, T., 2012. Landless farmers, sly opportunists, and manipulated voters: the squatters of Bukit Barisan Selatan National Park (Indonesia). Conserv. Soc. 10, 243–255.

Linkie, M., Dinata, Y., Nofrianto, A., LeaderWilliams, N., 2007. Patterns and perceptions of wildlife crop raiding in and around Kerinci Seblat National Park, Sumatra. Anim. Conserv. 10, 127–135.

Mangunjaya, F.M., McKay, J.E., 2012. Reviving and Islamic approach for Environmental Conservation in Indonesia. Worldviews 16, 286–305.

McCarthy, J.F., 2002. Turning in circles: district governance, illegal logging, and environmental decline in Sumatra, Indonesia. Soc. Nat. Resour. 15 (10), 867–886.

McGinnis, M.D., Ostrom, E., 2014. Social-ecological system framework: initial changes and continuing Challenges. Ecol. Soc. 19 (2), 30. http://dx.doi.org/10.5751/ES-06387-190230.

McKay, J.E., 2013. Lessons learned from a faith-based approach to conservation in West Sumatra, Indonesia. Asian J. Conserv. Biol. 2 (1), 84–85.

Measey, M., 2010. Indonesia: a vulnerable country in the face of climate change. Global Majority E: J. 1, 131–145.

Ostrom, E., 2007. A diagnostic approach for going beyond panaceas. Proc. Natl. Acad. Sci. 104 (39), 15181–15187. http://dx.doi.org/10.1073/pnas.0702288104.

Ostrom, E., 2009. A general framework for analyzing sustainability of social-ecological systems. Science, 419–422. http://dx.doi.org/10.1126/science.1172133.

Ostrom, E., Cox, M., 2010. Moving beyond panaceas: a multi-tiered diagnostic approach for social- ecological analysis. Environ. Conserv. 37, 451–463.

Silver, C., 2003. Do the Donors Have It Right?: Decentralization and Changing Local Governance in Indonesia. The Annals of Regional Science 37, 421–434.

Thorburn, C., 2000. Changing customary marine resource management practice and institutions: the case of Sasi Lola in the Kei Islands, Indonesia. World Dev. 28 (8), 1461–1479.

Wilcove, D.S., Giam, X., Edwards, D.P., Fisher, B., Koh, L.P., 2013. Navjot's nightmare revisited: logging, agriculture, and biodiversity in Southeast Asia. Trends Ecol. Evol. 9, 531–540.

Methodological Approaches in Natural Resource Management

R. Ullah
The University of Agriculture, Peshawar, Pakistan

2.1 INTRODUCTION

The history of the social sciences can be recounted with reference to major methodological shifts. An initial reliance on qualitative analysis gave way dramatically to quantification in the early to mid-20th century. When this transformation began, quantification largely meant statistical analysis of large-N data sets of public opinion surveys. The last third of the 20th century saw a surge in the use of formal models as well. Debates about the relative merits of qualitative, statistical, and formal methods contributed to several developments in the late 20th and early 21st centuries: refinements of quantitative methods that attempt to better match social conditions; the rise of formal models; greater appreciation for combining multiple methods; and the spread of postpositivist methods such as discourse analysis (Poteete et al., 2010).

There is growing recognition of the need for fresh ways of framing problems of sustainable consumption. Understanding consumption as integral to social practices provides a productive starting point (SPRG, 2012). Consumption as defined by Warde (2005) is "a moment in almost every practice"; key resources such as water, energy, and food are consumed as part of people's routine enactment of many different practices with direct and indirect implications for environmental sustainability. Focusing on the emergence, persistence, and disappearance of such practices and on the cultural norms, institutions, technologies, and infrastructures that constitute them can tell us much about the changing dynamics of sustainable and unsustainable consumption (SPRG, 2012).

Ecologists and governance scholars recognize and acknowledge the complex nature, both individually and as interacting systems, of natural resources and their governance (Mwangi and Wardell, 2012). Owing to the increasing awareness of the policy and management failure arising from ignoring scale and cross-scale dynamics in human-environment systems, efforts for understanding the structure and function of coupled social and ecological systems have been underway over several decades (Mwangi and Ostrom, 2008; Ostrom, 2009). However, there is still a major challenge for many researchers and practitioners on how to recognize and address cross-scale dynamics in space and over time in designing and implementing effective governance arrangements (Nagendra and Ostrom, 2012; Poteete, 2012).

2.2 MULTIMETHODS IN NATURAL RESOURCES MANAGEMENT (NRM)

Community-based natural resources management (CBNRM) involves both physical and socioeconomic systems. Interaction of these two systems influences both the viability of the natural resource and livelihood levels of the resource users. Conceptualization of the research approaches, methods, tools, and techniques for studying social processes and livelihoods outcomes need further research work and intellectual debate. In recent years, livelihoods for example, are increasingly conceptualized as partly the outcome of negotiations and bargaining between individuals with unequal power, even within households. Measuring the social capital, an interhousehold network of relationships for livelihoods has therefore become increasingly important in studying collective action in the NRM sector (Pokharel et al., 2002).

Government policy, regulations, officials' preconceptions, and attitude together with the relationships that the community has with them and vertical and horizontal linkages and relationships with other organizations affect how community forestry functions and collective action is promoted. The concept and methods of analyzing community structure, social and physical processes, methods of analyzing institutional linkages, and development interventions all take an important place in the methodology of studying collective action. The formal as well as informal relationships that community members have among themselves, with outsiders, and with the broader economic as well as the political process and the methods to study their relationships rather than study of organizations only pose a new conceptual challenge in the field of CBNRM.

The study of collective action to understand the relationships between people and natural resources without considering the outside organizations, economic, and sociopolitical forces that influence the collective action will therefore be incomplete and there is essentially a need for a combination of approaches, methods, and techniques. The structure and dynamics of community and natural resources, use of quantitative and qualitative information, assessment of the conditions of natural resources and the livelihoods outcomes, all these aspects have to be understood and captured in the study of collective action in CBNRM. Hence the application of research methods derived from both the natural and social sciences is necessary. In addition, the institutional research approaches and perspectives which are derived from common property literature, also need to be adapted in measuring collective action.

However, due to some conceptual problems in both natural and social science research approaches, there is a stronger need of a combination of various approaches to be applied in studying collective action for NRM. Collective action should not be seen in terms of community-resource relations only. Instead, CBNRM involves a number of various stakeholders at various levels, and the collective action among them also has to be recognized. The element of governance at all levels should be the focus of the study because many of the institutional conditions are derived from governance, which does affect the processes and outcomes of the collective action (Pokharel et al., 2002). Both qualitative and quantitative methods are entirely compatible and provide useful methodological complements. This provides the opportunity for a multidisciplinary team to work together.

Measuring collective action requires the study at macro, meso, and micro levels and their linkages. The linkage method has become useful to explain the influence of external factors on the ecological and social processes at the local level. As linkage research combines multilevel (international, national, regional, local) analysis and systematic comparison and longitudinal study (Kottak, 1999), this method enabled us to understand the link between the policies and practices at the micro level especially related to the process of inclusion and exclusion within resource users groups created under government and donor-funded programs. Uphoff (1998) identified actors and stakeholders at 10 different levels ranging from international, national, regional, district, subdistrict, locality, community/village, groups, to households and individuals. He argues that decision making and action can take place at any or all of these different levels that can influence the collective action and its outcomes.

2.3 MULTILEVEL ANALYSIS FOR NRM

Globalization and decentralization, and the multiscalar social and environmental changes associated with each, are two related processes that create a need for better understanding linkages across different spatial scales and governance levels (Berkes, 2008; Brondizio et al., 2009). For example, efforts at mitigating global climate change through Reducing Emissions from Deforestation and Forest Degradation (REDD+) and other initiatives are associated with global mechanisms such as carbon markets/credits and substantial financial transfers, while at the same time requiring the monitoring and conservation of forest resources at local levels (Angelsen, 2012). Both processes create different pressures at global, regional, national, subnational, and local levels, which may affect negatively or positively rules for resource access and use as well as incentives for sustainable use and management of forest resources. By increasing the number and type of actors, and the diversity of and asymmetries in interests, claims, and influence, these processes intensify the well-known problems of exclusion and substractibility that characterize common pool resources like forests, fisheries, and pastures, and may lead to a breakdown of previously effective arrangements for resource use and control. Ultimately, global, regional, national, and subnational influences are all mediated at the local level (Wardell and Lund, 2006).

Community participation and decentralization approaches have not been completely effective in linking different governance across levels. Considerable evidence suggests that they lead to elite capture and even negative resource outcomes (Poteete et al., 2010). The attributes of a resource, the attributes of users, and the institutional environment are factors worth considering in fostering collaboration across scales for NRM (Poteete and Ostrom, 2004). The form of cross-level interactions is also strongly influenced by the power relations inherent within them (Mwangi and Wardell, 2012). Different stakeholders use institutions and linkages to further their own interests (Lovell et al., 2002; Adger et al., 2005). Important elements of power include how decisions are negotiated, how and/or what trade-offs are made, and whether other actors are involved or not (Adger et al., 2005).

The state and its various agencies with authority at different levels are mentioned as crucial actors in effective NRM (Markelova and Mwangi, 2012). In watershed management the state can play a variety of roles at different scales (Swallow et al., 2001). It can facilitate the development and effectiveness of local organizations (local level), provide assistance through policy and financial support of group activities (municipal level), and promulgate favorable policies that help local organizations to be effective (national level).

Social scientists often examine relationships between variables measured at multiple levels of analysis. Multilevel methods consist of statistical procedures that are pertinent when

1. The observations that are being analyzed are correlated or clustered along spatial, nonspatial, or/and temporal dimensions; or
2. The causal processes are thought to operate simultaneously at more than one level; and/or
3. There is an intrinsic interest in describing the variability and heterogeneity in the population, over and above the focus on average relationships (Subramanian et al., 2003; Subramanian, 2004).

It is clear that individuals are organized within a nearly infinite number of levels of organization, from the individual up (eg, families, neighborhoods, counties, states, regions), from the individual down (eg, body organs, cellular matrices, DNA), and for overlapping units (eg, area of residence and work environment). Therefore, it is necessary that links should be made between these possible levels of analysis (McKinlay and Marceau, 2000).

2.4 METHODOLOGIES USED IN THE VOLUME

Keeping in mind the importance of using multimethods and multilevel analysis for NRM, all of the chapter authors of this volume explicitly used these techniques in their studies. The multimethods and multilevel analysis used in this volume are summarized below.

2.4.1 Multimethod and Multilevel Analysis of Forest Resources Management

2.4.1.1 Illegal Timber Felling and Comanagement of Protected Areas

Household participation in illegal timber felling in a protect area is modeled using logistic regression analysis. Sixteen contextual parameters are selected as independent variables and a principle component analysis with Kaiser normalization was performed to reduce the number of independent variables to factors. The number of independent variables is reduced to seven factors including household size and number of government subsidies, presence of local forest control, absence of high wealth possessions and collection of nontimber forest products, number of buffalo, possession of chainsaw, and involvement in hunting and trapping wildlife. The importance of timber felling activities to the household economy is also modeled using logistic regression analysis.

The important role played by each group of stakeholders in protected areas management is examined after decentralization. The varying degrees of decentralization in the districts are represented by D-I, D-II, D-III, and D-IV in Barisan I Nature Reserve. The multilevel and its resultant impacts on local government and *nagari* initiatives of protected areas comanagement has been described using exploratory methods. The perceptions of local people regarding decentralization and comanagement and the existing authority over the protected areas in the four decentralization zones are assessed using quantitative methods.

2.4.1.2 REDD+ Implementation

Multilevel analysis has been used to describe the diverse dilemma evolving in forest degradation and deforestation on the verge of implementing REDD+ in Indonesia. An overview of deforestation and forest degradation has been presented first, followed by institutional characteristics of deforestation. Local community rights over forest resources and conflicts arising in the forestry sector mainly due to unfair benefit sharing are discussed. REDD+ from equity, effectiveness, and efficiency perspectives has been qualitatively assessed.

2.4.2 Analysis of Water Resources and Watershed Management

2.4.2.1 Flood Forecasting and Early Warning System

Multimethod analysis was used in developing a flood forecasting and early warning system for the Air Dingin watershed. The flood forecasting model was developed using PCRaster software. While the qualitative method is used in developing the community-based flood early warning system, flood occurrence in Air Dingin watershed was modeled based on hydrologic characteristics of this region using PCRaster software, a prototype raster GIS for dynamic modeling. Static maps of maximum flood prone areas for 10, 25, 50, 100, and 200 years of return periods were forecasted. Prediction of rainfall for those return periods were calculated by Gumbel modification, Log Pearson Type III, and Iwai Kodoya method. Responses of the government and the community to flood occurrences were also described using exploratory research methods and recommendations were provided to effectively cope with the floods in the Air Dingin watershed.

2.4.2.2 Watershed and Livelihoods

The sustainable livelihood approach (SLA) was used to assess livelihood change and livelihood sustainability in the uplands of Lembang subwatershed, West Sumatra, Indonesia, in a changing NRM context. The SLA assessed the impact of internal and external factors of livelihood on household livelihood strategies and outcomes. The internal factors included capital assets (human, natural, physical, financial, and social) while external factors included vulnerability context and the transforming structures and process that affect a household's access to the internal factors. Vulnerability encompasses the risks, stresses, emergencies, and contingencies to which a household can be exposed, while structural context includes laws, policies, institutions, and governance that influence the access to assets and the scope for their application. The SLA is applied both for a better understanding of human-nature interrelationships, and for the integration of previous methodologies to assess livelihood sustainability as well as watershed management performance. In the SLA, livelihood sustainability is divided into four aspects: environmental sustainability, economic sustainability, social sustainability, and institutional sustainability. Livelihood sustainability of the Lembang subwatershed was assessed using the SLA in a changing NRM context. The major changes in the NRM context include (1) decentralization and restoration of the *nagari*, (2) changes in regulations on NRM, (3) economic crisis and macroeconomic structural adjustment, (4) biophysical changes, (5) changes in access to capital assets, and (6) changes in livelihood strategy.

2.4.2.3 Livelihood Security

Livelihood security of fishermen was assessed in Nagari Sungai Pisang, West Sumatra. Local people develop livelihood strategies to cope with fluctuating and unpredictable fish catches that threaten their livelihood security. These strategies include utilizing the kindness of nature, alternative productive activities, forecasting for unexpected future events, maintaining or enhancing social relationships, and risk-spreading mechanisms and productivity enhancement. The role of livelihood assets (ie, natural capital, social capital, human capital, financial capital, and physical capital), owned or accessed by the available institutions, was also assessed in enhancing capabilities to develop livelihood strategies. Moreover, the impact of culture and religion on the development of livelihood strategies was also investigated.

2.4.3 Multimethod Analysis of Land Use and Land-Use Change

2.4.3.1 Land-Use Change

Land-use changes along the Kuranji River Basin in Padang City (West Sumatra) were derived from a comparison of satellite images. The research first identified land-use according to a land utilization map and satellite images to study the changes. The satellite images used were derived from satellite image Landsat ETM +, in 2002, and UTM Zone47, southern hemisphere WGS84. Land Utilization Map of Padang City, scale1: 50,000, in 1994. The satellite image was classified based on land-use type using ERDAS software, and the land utilization map was manually digitized and grouped into land-use type using Map Info software. Both the satellite image and the map were compared to see the changes of percentage and total area of land cover. The linkage between the socioeconomic characteristics of the local communities and forest conditions were also described upstream of the basin. The impact of land-use change on river basin functions (ie, irrigation and municipal water supply) were also presented.

2.4.3.2 Sustainable Land-Use Practices

Observed facts, situations, and cases dealing with land and forest rehabilitation and conservation to support sustainable land-use practice were described and compared in two villages Nagari Paninggahan and Nagari Paru. The role of incentive factors in sustainable land-use practices (implementation and development of rehabilitation and conservation initiatives) were also compared in the two villages. Similarities and differences of incentive factors in Paru and Paninggahan villages were described in terms of knowledge and awareness, benefits, material, financial, and technical incentives, ease of access to services and infrastructure, land tenure system, and policies and institutional factors. The consequences of participation and nonparticipation in rehabilitation and conservation initiatives were also described in the two villages.

2.4.4 Property Rights, Gender Inequality, and Sustainability of Natural Resources

2.4.4.1 Community-Based Property Rights

Community-based property rights of the local community of Sungai Pisang, West Sumatra, was observed/explored. The main aim of the research was to explore how the local community was optimizing the utilization rights of the small islands as their community-based property rights after leasing, through observed facts and information from the key informants. The utilization rights implementation in Sikuai and Pasumpahan islands was explained based on the information from key informants. The research was also aimed at identifying the problems during utilization rights implementation and the problem-solving mechanisms implemented/used.

2.4.4.2 Gender Inequality

To present observed facts and situations related to gender inequality of a fishing family in Padang Pariaman District and female participation in a rural water supply project in Solok, an exploratory method was used to provide richer insights into a given situation. These methods were directed to highlight genderwise access to family assets and resources. Female participation at every stage (project initiation and decision making, project construction, operation and maintenance, monitoring and evaluation, and project sustainability) of the rural water supply project was also explored. Factors influencing gender inequality and female participation were also figured out.

2.4.4.3 Hexagon Framework

In addition to a pentagon framework that categorized various important aspects to identify how far a development process can be interfered, a hexagon framework was also used to describe the impact of religious values and cultural practices on sustainability of NRM. The key Islamic values [Tawheed (unity), Mizan (equity), Ad'l (justice). Khalifa (leadership), and use not abuse] were expressed against the indictors of sustainable NRM (ie, reference, restrain, redistribution, respect, and responsibility).

2.5 CONCLUSION

There should be an appropriate mix of local and state institutions in governance arrangements for NRM with strong support by central state authorities. There is a need for support from state institutions for the formation or strengthening of these local institutions where they are nonexistent or weak, and to mediate conflicts and enforce resource use agreements worked out by the different local groups (Ostrom, 1990, 1995). It is impossible to capture and account for the true complexity of human-resource interactions without disaggregating by scale and looking at cross-level linkages. There is wide agreement as well that the type of institution must be matched to the scale of the resource while fostering accountable cross-scale linkages among multiple actors. There is value in both the top-down approach of decentralization and the bottom-up approach of community participation. Yet both approaches are susceptible to some common problems, especially elite capture at different levels, which can ultimately hinder healthy cross-scale linkages (Mwangi and Wardell, 2012).

Thus multilevel governance devotes attention to the links between humans and their environment, which may occur vertically (ie, from local to global) or horizontally (at the same level), as well as to contestation and learning among parties with a stake in forests and other natural resources (Armitage, 2008; Berkes, 2008; Brondizio et al., 2009). It provides a framework for analysis and scope to address complex multiscale/level problems related to NRM (Termeer et al., 2010), albeit often at the intersection of different epistemological traditions (Mwangi and Wardell, 2012).

References

Adger, W.N., Brown, K., Tompkins, E.L., 2005. The political economy of cross-scale networks in resource co-management. Ecol. Soc. 10 (2), 9. Available online, http://www.ecologyandsociety.org/vol10/iss2/art9/.

Angelsen, A. (Ed.), 2012. Analyzing REDD+ Opportunities and Choices. Center for International Forestry Research, Bogor.

Armitage, D., 2008. Governance and the commons in a multi-level world. Int. J. Commons 2 (1), 7–32.

Berkes, F., 2008. Commons in a multi-level world. Int. J. Commons 2 (1), 1–6.

Brondizio, E.S., Ostrom, E., Young, O.R., 2009. Connectivity and the governance of multilevel social-ecological systems: the role of social capital. Annu. Rev. Environ. Resour. 34, 253–278.

Kottak, C.P., 1999. The new ecological anthropology. Am. Anthropol. 101 (1), 23–35.

Lovell, C., Mandondo, A., Moriarty, P., 2002. The question of scale in integrated natural resource management. Conserv. Ecol. 5 (2), 25. Available online, http://www.consecol.org/vol5/iss2/art25/.

Markelova, H., Mwangi, E., 2012. Multilevel governance and cross-scale coordination for natural resource management: Lessons from current research. CIFOR, Bogor, Indonesia. Available at: http://wealthofthecommons.org/essay/multilevel-governanceand-cross-scale-coordination-natural-resource-management-lessons-current.

McKinlay, J.B., Marceau, L.D., 2000. To boldly go. Am. J. Public Health 90 (1), 25–33.

Mwangi, E., Ostrom, E., 2008. A century of institutions and ecology in east Africa's Rangelands: linking institutional robustness with the ecological resilience of Kenya's Maasailand. In: Beckmann, V., Padmanabhan, M. (Eds.), Institutions and Sustainability. Political Economy of Agriculture and the Environment. Essays in Honor of Konrad Hagedorn. Springer, Dordrecht, pp. 195–222.

Mwangi, E., Wardell, A., 2012. Multi-level governance of forest resources. Int. J. Commons 6 (2), 79–103.

Nagendra, H., Ostrom, E., 2012. Polycentric governance of multifunctional forested landscapes. Int. J. Commons 6 (2), 104–133.

Ostrom, E., 1990. Governing the Commons. The Evolution of Institutions for Collective Action. Cambridge University Press, Cambridge.

Ostrom, E., 1995. Designing complexity to govern complexity. In: Hanna, S., Munasinghe, M. (Eds.), Property Rights and the Environment: Social and Ecological Issues. The Beijer International Institute of Ecological Economics and the World Bank, Washington, DC, pp. 33–45.

Ostrom, E., 2009. A general framework for analyzing sustainability of social-ecological systems. Science 325, 419–422.

Pokharel, B.K., Ojha, H.R., Paudel, K., 2002. Methods of studying collective action in natural resource management: the case of community forestry in Nepal. In: Paper presented at the CAPRi Workshop on Methods for Studying Collective Action, February 25–March 1, 2002, in Nyeri, Kenya.

Poteete, A., 2012. Levels, scales, linkages, and other 'multiples' affecting natural resources. Int. J. Commons 6 (2), 134–150. http://dx.doi.org/10.18352/ijc.318.

Poteete, A., Ostrom, E., 2004. An Institutional Approach to the Study of Forest Resources. IFRI, Bloomington. Manuscript available at http://www.indiana.edu/~workshop/papers/W01I-8.pdf.

Poteete, A., Janssen, M.A., Ostrom, E., 2010. Working Together: Collective Action, the Commons, and Multiple Methods in Practice. Princeton University Press, Princeton, NJ.

SPRG, 2012. Researching Social Practice and Sustainability: Puzzles and Challenges. Sustainable Practices Research Group working paper 2. Available online on http://www.sprg.ac.uk/uploads/practices-and-methodological-challenges.pdf.

Subramanian, S., 2004. The relevance of multilevel statistical methods for identifying causal neighborhood effects. Soc. Sci. Med. 58 (10), 1961–1967.

Subramanian, S., Jones, K., et al., 2003. Multilevel methods for public health research. In: Kawachi, I., Berkman, L. (Eds.), Neighborhoods and Health. Oxford Press, New York, pp. 65–111.

Swallow, B.M., Garrity, D., van Noordwijk, M., 2001. The effects of scales, flows and filters on property rights and collective action in watershed management. Water Policy 3, 457–474.

Termeer, C.J.A.M., Dewulf, A., van Lieshout, M., 2010. Disentangling scale approaches in governance research: comparing monocentric, multilevel, and adaptive governance. Ecol. Soc. 15 (4), 29.

Uphoff, N., 1998. Community based natural resource management: connecting micro and macro processes, and people with their environments. In: Plenary on International Workshop on Community Based Natural Resource Management, Washington, DC. http://www.worldbank.org/wbi/conatrem/uphoff-paper.htm.

Warde, A., 2005. Consumption and theories of practice. J. Consum. Cult. 5 (2), 131–154.

Wardell, D.A., Lund, C., 2006. Governing access to forests in northern Ghana: micro-politics and the rents of non-enforcement. World Dev. 34 (11), 1887–1906.

Livelihood Change and Livelihood Sustainability in the Uplands of Lembang Subwatershed, West Sumatra Province of Indonesia, in a Changing Natural Resources Management Context

Mahdi*, G. Shivakoti[†,‡], D. Schmidt-Vogt[†]

*Andalas University, Padang, Indonesia [†]Asian Institute of Technology, Bangkok, Thailand [‡]The University of Tokyo, Tokyo, Japan

3.1 INTRODUCTION

Improving the livelihoods of local people has received growing attention during last two decades, and is one of the main goals of watershed management. The prime engine to achieve this goal, according to Mitchell (2002, 2005), is integrated watershed management (IWM). Livelihoods have been incorporated in IWM by applying the livelihood concept. This can be done in two ways. First, it is utilized as a perspective to view the livelihood of local people and their strategies on resources use in the context of human-nature interrelationships in a watershed (Scoones, 1998; Arnold, 1998; Torras, 1999; WRI, 2001; Dewi et al., 2005; WRI, 2005; Vedeld et al., 2007). Second, the sustainable livelihood framework (SLF), which has been developed by international agencies such as the UK Department for International Development (DFID), CARE, Oxfam, and the United Nations Development Program (Carney et al., 1999), is applied for the identification of indicators to measure watershed management performance in an IWM (Campbell et al., 2001; Shivakoti and Shrestha, 2005a).

The problems of watershed management become more complex as livelihoods change over time due to altering external and internal factors. Livelihood change affects the environment and vice versa (Dupar and Badenoch, 2002; WRI, 2001), ultimately leading to effects on livelihood sustainability as well as watershed sustainability. Little attention, however, has been paid to livelihood change and its integration in a watershed. This chapter argues that policy makers and resources managers should take these issues into consideration for an adaptive approach to IWM, as the adaptive capacity in an IWM is the main strategy to achieve better performance of natural resources management (NRM) (Armitage, 2005; Armitage et al., 2008).

Changing the external factors of livelihood is in this chapter is referred to as changing the NRM context. Changing the internal factors, on the other hand, refers to the change in access to livelihood capital assets including human, natural, physical, financial, and social capital. In Indonesia, the NRM context has changed dramatically during the last decade. The social and political situation has changed rapidly after the collapse of the centralized and authoritarian Suharto regime due to an economic crisis that hit Indonesia in the mid-1990s. Following this collapse, decentralization was enacted in 2000 in an effort to respond to the demand for political, administrative, and economic reform, including decentralization of NRM (Resosudarmo, 2002). The government of Indonesia (GoI) issued and implemented Law No. 22/1999 regarding the decentralization of regional governments. In addition, Forestry Law No. 41/1999 and Water Resources Law No. 7/2004 replaced older laws. These new laws recognize the role of traditional rules and regulations in NRM. In addition, the economic crisis caused the GoI to adjust its macroeconomic policy by devaluating the exchange rate of the rupiah, by promoting economic liberalization, by bringing about changes in government expenditure, and by adjusting privatization and interest rates (San et al., 2000). These policy changes have increased the social and environmental cost of reliance on resource-based export growth (Gellert, 2005). In addition to changes on the economic and political level, changes on the natural level over the last decade, such as the more pronounced and more frequent occurrence of La Niña and El Niño events, have led to an increase in natural disasters (Irianto et al., 2004).

The changes of context affect the internal factors of livelihoods and livelihood strategies at the household level (O'Connor, 2004; Scoones, 1998). Our concern in this chapter is to understand whether or not these changes lead to sustainable livelihood, and how changes differ among the various groups in a heterogeneous community. To address this concern, we analyze the pattern of livelihood change of upland inhabitants of the Lembang subwatershed of West Sumatra, Indonesia, by measuring access to capital assets and by analyzing livelihood strategies. Changes are investigated at two separate points in time, 1996 and 2006. Following that, we assess the effect of both external and internal factors as well as the effect of livelihood strategies on livelihood sustainability.

3.2 LIVELIHOOD AND IWM

The concept of livelihood for the poor emerged out of the Brundtland Commission's sustainability report, which argued for balanced development with equal emphasis on ecological and social aspects (WCED, 1987). The report opened up a new way for poverty reduction, especially because the previous approach of integrated rural development has had little or no impact on poverty reduction (Ashley and Carney, 1999; Chambers, 1995; Bebbington, 1999).

The SLF, developed by Chambers and Conway (1992), and later by DFID (Fig. 3.1), formulates schematically the interaction between the internal and external factors of livelihood, which

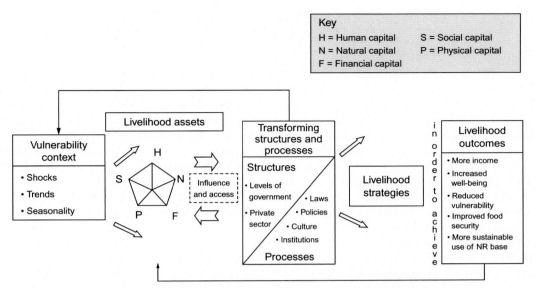

FIG. 3.1 Sustainable livelihood approach DFID, 1999.

determine household livelihood strategies and outcomes (Koeberlein, 2003). Internal factors are the five capital assets (human, natural, physical, financial, and social) that a household has access to. Access to these assets is influenced by external factors. A human capital asset is, for instance, the amount and quality of labor available. The natural capital asset comprises the natural resources, from which a livelihood can be derived. The physical capital asset contains the basic infrastructures and means of production, and the financial capital asset the financial resources needed to support a livelihood. The social capital asset indicates the involvement of household in social activities and networks for both political and economic purposes.

External factors are the vulnerability context, and the transforming structures and process. Vulnerability comprises the risks, stresses, emergencies, and contingencies to which a household can be exposed. The access to assets and the scope for their application is influenced by the structural context that encompasses laws, policies, institutions, and governance. The structural context also has an impact on livelihood strategies.

Household livelihood sustainability refers to the ability of a household to deal with shocks and stresses, and to maintain or enhance its capabilities and assets without jeopardizing the natural resources base (Chambers and Conway, 1992). In this framework, livelihood sustainability is divided into four aspects (DFID, 1999). First is environmental sustainability, which is achieved when natural resources for livelihood support are conserved or enhanced over time. Second is economic sustainability, which is achieved when a given level of income as well as expenditure can be maintained or increased over time. Third is social sustainability, which is achieved when social exclusion is reduced, and social equity enhanced. Fourth is institutional sustainability, which is achieved when institutions for sociopolitical and resources governance have the capacity to continue and perform their functions over the long term. Livelihood sustainability indicators can also serve as indicators for assessing IWM performance (Campbell et al., 2001).

3.3 METHODS

3.3.1 Research Framework

In this paper, livelihood changes and livelihood sustainability are examined by applying the SLF in response to a changing NRM context within a watershed. For this purpose, the SLF is applied both for a better understanding of human-nature interrelationships, and for the integration of methodologies developed by Campbell et al. (2001) and Shivakoti and Shrestha (2005a,b) to assess livelihood sustainability as well as watershed management performance. Fig. 3.2 depicts a research framework that shows how the changing NRM context correlates with the changing access to capital assets. The changing context and capital assets, then, jointly influence the changing livelihood strategies. Changing context, capital assets, and livelihood strategies then affect in unison the livelihood sustainability. Livelihood sustainability is also a reflection of watershed management performance.

To integrate this framework into our research, we adapt the SLF in three consecutive steps. First, we develop and apply a quantitative technique for livelihood changes measurement at two separate points in time, 1996 and 2006, by identifying and formulating indicators for access to capital assets. Identification and formulation of indicators are derived from the variables that would be affected by changing the NRM context. Table 3.1 recapitulates the capital assets, their building variables, and their index measurements, and how access to them is affected by a changing NRM context. Second, the responses of households to changes in access to capital assets as well as changes in the NRM context are deduced from changes in their livelihood strategies. Third, the trends of change in internal and external factors, and in livelihood strategies, are linked to livelihood sustainability. The sustainability assessment is done qualitatively by correlating changes and impacts with indicators of sustainability as formulated by the DFID (1999).

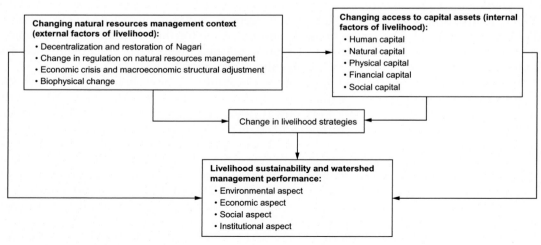

FIG. 3.2   The research framework.

TABLE 3.1 Livelihood's Capital Asset, the Building Variables and the Index to Measure Their Change

No	Capital Asset and Its Building Variables	Index	External Factors That Affect the Change in Access to Capital Asset
1	Human capital	Average of labor and education index	
	Labor	Ratio of workforce in a household	• Family members dynamics
	Education	Ratio of family members with senior high school education or above	• Structural adjustment
2	Natural capital	Average of access to forest resources and access to land index	
	Forest resources	Percentage of family income from timber and nontimber forest products (NTFP)	• Issue new forest law • Decentralization and restoration of *nagari*
	Land	Security of land ownership. Private land is valued 1; lineage land, 0.5; and rent/sharecropper, 0.2	• Issue new forest law • Decentralization and restoration of *nagari*
		Quality of land. Paddy field is valued 1; dry land, 0.5; and other, 0.2	• Issue new forest law • Decentralization and restoration of *nagari* • Biophysical change
3	Physical capital	Average of access to irrigation and road infrastructures, farm input accessibility, and processing index	
	Irrigation infrastructure	Percentage of farmer's paddy field receiving irrigation water	• Structural adjustment • Issue new water law
	Road infrastructure	Comparison of walking time from nearest agricultural land in a *nagari* to service road vs. average walking time of respondent's agricultural land to service road	• Structural adjustment
	Farm input	Average percentage of households applying chemical or organic fertilizer, practicing integrated pest management, and applying high-yield variety of seed in their agricultural practices	• Structural adjustment
	Processing	Average percentage of processed commodities sold, by household	• Structural adjustment

Continued

I. LIVELIHOOD DEPENDENCE, RIGHTS AND ACCESS TO NATURAL RESOURCES

TABLE 3.1   Livelihood's Capital Asset, the Building Variables and the Index to Measure Their Change—cont'd

No	Capital Asset and Its Building Variables	Index	External Factors That Affect the Change in Access to Capital Asset
4	Financial capital	Average of access to cash income, credit, and subsidy-tax index	
	Cash income	Percentage of cash income	• Structural adjustment
	Credit	Ratio of credit received by household to the value of household's immobile private assets	• Structural adjustment
	Subsidy-tax	Amount of received subsidy minus taxes paid divided by total cash income	• Structural adjustment
5	Social capital	Average of access equity and institutional participation index	
	Equity	Ratio of household income per capita to *nagari* income per capita	• Decentralization and restoration of *nagari* • Structural adjustment • Biophysical change
	Institutional participation	Percentage of households or heads of family participating in forest and water organization activities during a year	• Decentralization and restoration of *nagari* • New forest and water law

This research opens up a new dimension of measuring livelihood change both quantitatively and qualitatively. Quantitative measurement in the form of indexing access to capital assets, helps us to follow the pattern of change in access to capital assets. Which assets are accessed more, and who is better off, can be determined with higher accuracy than by using qualitative methods only, which was commonly done in livelihood analysis. Qualitative analysis, on the other hand, captures information of the kind that lies beyond the scope of quantitative measurement. The latter supports and strengthens the former technique.

However, one should be aware of three weaknesses of this method. First, indexing of access to capital assets cannot include all building variables. This may lead to bias. This weakness can be mitigated, however, by selecting those building variables that can reflect best the effects of the changing NRM context. Second, it is not possible to determine which context has the greater influence on change in access to capital assets, on livelihood strategies, and on livelihood sustainability. The framework can only predict that all changes of context collaboratively affect the changes of livelihood and of livelihood sustainability. Third, measuring at two separate points in time can only capture the indications within these points in time, missing out on the livelihood dynamics between them.

3.3.2 Rapid Rural Appraisal and Household Survey

Rapid rural appraisal (RRA) and household surveys were carried out in three *nagari*s (villages) within the upland of the Lembang subwatershed: *Nagari* Selayo Tanang Bukik Sileh, *Nagari* Koto Laweh, and *Nagari* Dilam. RRAs were carried out by interviewing key informants and organizing focus group discussions to learn about conflicts and local institutional changes during the last decade. Household surveys, on the other hand, were carried out to obtain data at the household level concerning changes in access to capital assets and livelihood strategies. Household samples were taken randomly from three groups: low income, middle income, and high income. Grouping of the population was based on the latest monthly household income per capita records kept in these three *nagaris*. Low income meant less than Rp 250,000 per month, middle income was in the range of Rp 250,000–550,000, and high income was more than Rp 550,000. Table 3.2 shows the household sample characteristics for each income group.

3.3.3 Data Analysis

Two statistical analyses were carried out. First, a *t*-test was conducted to examine the significances of differences in access to capital assets at two separate points in time, 1996 and 2006. Second, a one-way analysis of variance (ANOVA) test was conducted to examine the significance of differences in access to capital assets among households from the three different income groups. Then, an asset pentagon was drawn to show the pattern of change of access to capital assets between two separate points in time.

Changes in livelihood strategies and livelihood sustainability were analyzed qualitatively. Changes in livelihood strategies were identified by assessing the frequency of respondents' answers concerning their response to a changing NRM context and the changes in access to capital assets. Livelihood sustainability assessment was based on trends of change in the NRM context and their impact on change in access to capital assets, and on change in livelihood strategies. These parameters were then grouped according to the four aspects of sustainability: environmental, economic, social, and institutional. Trends and impacts were assessed with respect to their positive or negative effects on livelihood and watershed sustainability using the DFID's livelihood sustainability indicators. Livelihood sustainability indicators were also the basis for assessing watershed management performance.

TABLE 3.2   Household Sample Characteristics

No	Household Characteristics	Income Group			Total
		Low	Middle	High	
1	Household number (*N*)	94	41	25	160
2	Average age of the head of household (year)	45.85	45.73	49.40	46.38
3	Average monthly household income per capita (Rupiah)[a]	149,438.86	349,378.05	994,963.47	332,786.50
4	Average household size	5.27	3.88	3.40	4.62
5	Average years of formal education	6.45	7.32	6.60	6.69

[a] *During the study period 1 US dollar was equal to 9000 Indonesian Rupiah.*

3.4 STUDY SITE OVERVIEW

Research was carried out in the uplands of the Lembang subwatershed. Soils are fragile due to steep slopes, and most people depend for their livelihood on dry land agriculture and forest resources. Lembang subwatershed is located in the southern part of the Sumani watershed, which is the most important watershed in the central part of West Sumatra. It is under Solok District administration (see Fig. 3.3). Talang Mountain, the most active volcanic mountain in Sumatra, is also situated in the study area. In the research site, altitude ranges from 900 to 1700 m above sea level. The number of rainy days in a year ranges between 34 and 212, and average annual rainfall is 7768 mm. The lowest rainfall is usually recorded in Jul., the highest in the time from Nov. to Feb. Annual average temperature ranges between 12.5°C and 24.60°C. With 352 persons per square kilometer, population density in this subwatershed is the highest in the entire Sumani watershed. Annual population growth was around 1.3% during the last 5 years (Statistics of West Sumatra Province, 2005).

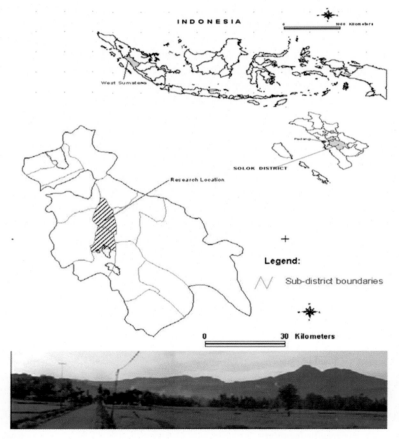

FIG. 3.3   Research site.

3.5 CHANGES IN NRM CONTEXT

3.5.1 Decentralization and Restoration of the Nagari

Decentralization has been implemented in Indonesia since 2000. In response to this policy, West Sumatra Province has formulated a provincial regulation to give a legal basis for restoring the *nagari* institution, and replacing the *desa* system of village administration, which used to be a common framework throughout Indonesia. *Nagari* is the lowest level political unit of the Minangkabau ethnic group, to which almost all of people in the study site belong. The Minangkabau practice the matrilineal system. A *nagari* is composed of several neighboring hamlets. Each hamlet represents a clan (*suku*) led by a *Datuk*. A clan has several lineages (*kaum*), each of which is led by a *Mamak Barih* who is selected from among the *Mamak Rumah*, the representatives of an extended family (*paruik*). *Nagari* has a democratic, autonomous, and informal structure with the clan and hamlet leaders placed on top (Naim, 1984). Because *nagari*s are independent institutions, each *nagari* has its own rules and laws.

Nagari is a promising local institution for NRM because decision making is made by local people. This is different from the approach criticized by Li (2002) by which outsiders' facilitation of community-based natural resources management (CBNRM) led to failure. Restoration of the *nagari* is meant to reestablish and formalize customary rules on NRM, and to encourage wider participation. Local leaders are now trying to reinstall their authority over forestland management, which had previously been weakened by handing over *nagari*s' ownership and land utilization rights to individuals and groups. Each *nagari* has written and formalized its customary rules. With respect to forest management, *nagari*s recently have made their own rules on forest products withdrawal. *Nagari*s also enforce these rules, and carry out evaluation and monitoring (Yonariza and Shivakoti, 2008). *Nagari* Kinari of Solok District wrote and enacted formally the customary rule on fisheries in *nagari*'s river in 2001 by issuing *Nagari* Kinari Regulation No. 04/2001. The regulation emphasizes that the rivers within the *nagari* should be used only with permission from the *nagari* government, otherwise the activities will be categorized illegal and the *nagari* has the right to impose a ban or a fine.

However, due to the fact that the decentralization act has been enacted only recently, the *nagari* is still not sufficiently capable to tackle all NRM problems. A number of conflicts over NRM issues have emerged within a *nagari*, between *nagari*s, as well as between *nagari*s and the local government. Within *nagari*s, conflicts have arisen because of overlap of ownership and the willingness of local leaders to reinstall their authority on *nagari*'s land, and because of unclear borders and overlapping claims of lineages and individuals. Uncertainty of rules during the early phases of restoring *nagari* exacerbate the conflict among people within the *nagari*, who respond to uncertainty by occupying more land and collecting more nontimber forest products (NTFPs) from the *nagari*'s forest. Conflicts among neighboring *nagari*s have arisen because of unclear boundaries among *nagari*s, and because there is still a lack of rules for mediation in conflict. Conflicts between *nagari*s and the local government are over irrigation management. *Nagari*s claim that, based on customary rule, the irrigation canals within their territory are owned by them. This claim, however, is not recognized in formal regulations.

With respect to livelihood, the restoration of *nagari* rule affects the access to capital assets in two ways. First, uncertainty concerning rules of NRM during the early stages of restoring the *nagari* has encouraged people to occupy more land, and to extract more NTFPs from the *nagari*'s forest. At the same time, this uncertainty has discouraged local people from participating in the *nagari*'s activities. Formalization of customary rule in NRM could solve this problem of uncertainty, and encourage participation for sustainable resources management. Second, conflict over resources utilization causes insecurity of access to capital assets.

3.5.2 Changes in Regulations on NRM

In 1999, the GoI issued Forestry Law No. 41/1999 to replace the old one. The new law provides wider scope for involving all stakeholders. It acknowledges local customary laws with respect to ecological and social aspects of forest management (MoF, 2003). Although it has been criticized that customary forest is still held as national forest estate under the control of the forestry department (Li, 2002), the new law gives wider space for local people and their institutions to manage their customary forest. Along with the restoration of the *nagari*, this law provides an opportunity for the people of West Sumatra to write and formalize their own customary laws regarding forestland and forest resources. *Nagari* Sungai Kamuyang of 50 Kota District, for instance, has issued a *nagari* regulation in 2003 regarding the utilization of *nagari*-owned land, including forest resources.

Furthermore, in 2004, the GoI also issued new Law No. 7/2004 on water resources, replacing Law No. 11/1974. The new law spells out the responsibilities of provincial and district governments as well as of farmers with respect to water management and, furthermore, invites participation from private enterprises to manage and supply drinking water. In addition, the new law emphasizes water resource conservation and protection. In relation to this new law, some *nagari*s in West Sumatra have formulated a *nagari* regulation on water resources management within their territory. In the study area, *Nagari* Kinari enacted a regulation on irrigation management in 2004, which claims that irrigation canals within its territory are owned and should be managed by the *nagari*.

These two laws provide scope for the local administration as well as for people at the local level to play a substantial role in forest and water management. As changes of context, they influence household access to capital assets including access to forest resources, forestland, and irrigation infrastructure. They also influence participation in water management institution. The indexes, which are presented in Table 3.1, show the effect of these new laws on livelihood change both directly and indirectly.

3.5.3 Economic Crisis and Macroeconomic Structural Adjustment

To respond to the economic crises of 1997, the GoI made structural adjustments in five areas: currency devaluation, price and economic liberalization, change in public expenditure, privatization, and interest rate adjustment (San et al., 2000). Before the economic crises, the exchange rate was Rp 2500 per US dollar; the rate increased to Rp 9000 and then to Rp 10,000 per US dollar after the economic crises. The currency devaluation caused an increase in the price of tradable goods and a decrease in the price of nontradable goods (San et al., 2000).

The GoI also liberalized the market by removing trading barriers and privatizing government-owned enterprises such as *Badan Urusan Logistik* (BULOG; Indonesian national logistics agency), the agency authorized to maintain rice price stability and ensure availability of food supplies at an affordable price (Robinson et al., 1998). These measures had a twofold effect on the livelihoods of poor rural households. First, the prices for domestic agricultural products fell because cheap and subsidized foreign products flooded Indonesia's market as import barriers were removed. This led to a reduction in the growth of the agricultural sector. Feridhanusetyawan and Pangestu (2003) calculated that the output of Indonesian paddy rice declined by 0.9% due to unilateral liberalization of trading. Second, the Indonesian government was under pressure by international trading communities to remove subsidies for agricultural inputs. This led to an increase in production costs, further eroding farmer competitiveness. Haryati and Aji (2005) reported that the price of fertilize has a negative imp act on rice productivity because farmers tend to reduce fertilizer usage when the price of fertilizer increases. The price of fertilizer almost doubled after the economic crisis. In the study site, the price of urea fertilizer, for instance, was Rp 1000 kg^{-1} in 1996, and increased to Rp 2000 kg^{-1} in 2006.

The GoI also adjusted its expenditures as part of a crisis and recovery program. While government expenditures for construction and maintenance of infrastructure were reduced, subsidies were provided to help poor households face the difficulties of increasing prices for staple food, health services, and education. For instance, *Jaring Pengaman Sosial* (JPS; safety nets program), a program funded by the World Bank, provided cash to help poor households face the impact of the economic crisis. A number of projects related to agriculture, education, health, employment, and other welfare-improvement activities have been carried out through the JPS program. Recently, a poor household was given Rp 300,000 in cash every 3 months, as well as 10 kg of subsidized rice per month at 75% of the market price by presenting a "kartu miskin" (poor card) to the *nagari*'s officials. The *kartu miskins* are issued by the *nagari* administration to categorized poor households. The card also needs to be shown when they need health care services for free in *Puskesmas* (health service centers) and when they make a request for a scholarship and an allowance for their childrens' education.

With respect to livelihood, measures such as these affect households' access to capital assets, especially access to irrigation infrastructure, road infrastructure, farm input and processing, cash income, credit and subsidy-tax, education, and labor. The index measurements for these assets reflect the effect of structural adjustment on livelihood change.

3.5.4 Biophysical Changes

Biophysical changes are related to land-use changes and natural hazards. To identify these changes, land use from different periods of time, as recorded in secondary sources, is presented in Table 3.3 for the Lembang subwatershed. Over time, there has been a decrease in the area of forest and shifting cultivation, and an increase in the area of bush land, settlement, paddy field, and degraded land. Extensive use of inorganic fertilizer and pesticides has negatively affected soil quality. Deforestation and intensive land tillage for vegetables cultivation have resulted in heavy erosion. Degradation of resources have forced local people to look for alternative means for livelihood support.

The main problem is soil erosion due to intensive tillage on a steep slope. Istijono (2006) reported that the annual average soil erosion of 154 tons within this subwatershed is the second highest within the Sumani watershed. Soil erosion reduces soil fertility, and has a direct impact on agricultural productivity and ultimately on local livelihoods.

A most powerful natural hazard affected the study site when Talang Mountain erupted in Apr. 2005. This forced the local people to stay at a refugee camp for more than 1 week.

TABLE 3.3   Lembang Subwatershed Land-Use Change 1890, 1976, 1993, and 2004

Land Use	1890		1976			1993			2004		
	Ha	%	Ha	%	% Cumm. Change	Ha	%	% Cumm. Change	Ha	%	% Cumm. Change
Forest	4142.67	24.25	1379.29	8.07	−66.71	1322.36	7.74	−68.08	509.71	2.98	−87.70
Shifting cultivation	1628.30	9.53	661.47	3.87	−59.38	5666.33	33.16	248.00	940.94	5.51	−42.21
Bush	2851.06	16.69	1488.34	8.71	47.79	443.90	2.60	−84.43	3158.78	18.49	10.79
Settlement	713.91	4.18	4750.92	27.81	565.48	1983.30	11.61	177.80	3907.74	22.87	447.37
Paddy field	423.40	2.48	472.34	2.76	11.56	522.57	3.06	23.42	597.96	3.50	41.23
Degraded	7326.84	42.88	8333.82	48.78	13.84	7147.72	41.83	−2.44	7971.06	46.65	8.79
Total	17086.18	100.00	17086.18	100.00		17086.18	100.00		17086.19	100.00	

From Istijono, B., 2006. Konservasi Daerah Aliran Sungai dan Pendapatan Petani: Studi Tentang Integrasi Pegelolaan Daerah Aliran Sungai. Studi Kasus DAS Sumani Kabupaten Solo/ Kota Solok, Sumatera Barat (Unpublished-Dissertation). Universitas Andalas, Padang.

They were also forced to sell their cattle at extremely low prices. The eruption has also had a negative effect on agricultural productivity as the ashes from the volcanic eruption contained sulfur, which affects soil fertility. Recently, this area has once again been under high alert.

3.6 UPLAND PEOPLE'S LIVELIHOOD CHANGE

3.6.1 Changes in Access to Capital Assets

Access to natural and physical capital assets increased significantly during the last 10 years, while access to human and financial capital has increased only slightly. During the same period, however, access to social capital decreased slightly as can be seen in Fig. 3.4. While access to labor has decreased only slightly, access to education increased significantly. This has resulted, however, in only an insignificant change in the human capital index. Most of the building variables of the natural capital and physical capital index have increased. Financial capital access has increased slightly as a compound effect of access to cash income and the subsidy-tax index having increased, and access to credit having decreased significantly. Access to social capital has decreased slightly due to a marginal decrease in the equity index and a significant decrease of institutional participation (see Table 3.4).

With respect to access to labor, there was a significant difference between income groups. High income groups had more labor in 1996, and less labor in 2006. The access index to labor for low- and middle-income families, however, has increased marginally during the same

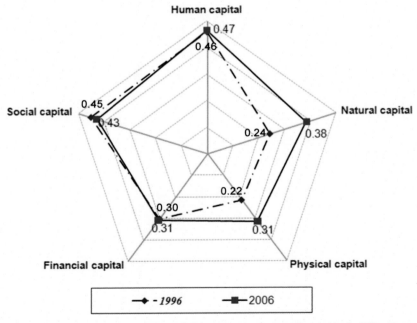

FIG. 3.4 Change of upland people's access to capital assets in Lembang subwatershed of West Sumatra, Indonesia 1996–2006.

TABLE 3.4 Capital Asset Access Index and Its Building Variables by Income Groups in 1996 and 2006

No	Capital Asset and Its Building Variable	Year	Low Income Average Index	Low Income t-Test	Middle Income Average Index	Middle Income t-Test	High Income Average Index	High Income t-Test	One-Way ANOVA	Total	t-Test
I	Human capital asset	1996	0.41	1.9904[a]	0.52	2.0011[a]	0.55	0.4460	8.3849[a]	0.46	0.2776
		2006	0.46		0.44		0.54		1.9755[a]	0.47	
	Labor	1996	0.73	0.2500	0.84	-2.6047[a]	0.93	-1.9908	9.8355[a]	0.79	-1.9117
		2006	0.74		0.70		0.83		2.6067[a]	0.74	
	Education	1996	0.09	3.5101[a]	0.19	0.1687	0.18	1.5285	2.3892[a]	0.13	3.3213[a]
		2006	0.17		0.19		0.24		0.7934	0.19	
II	Natural capital asset	1996	0.25	5.9512[a]	0.22	4.9054[a]	0.27	3.9920[a]	0.5300	0.24	8.6477[a]
		2006	0.37		0.38		0.43		1.6711	0.38	
	Forest resource	1996	0.09	3.3300[a]	0.15	-1.4100	0.09	0.8631	3.0006[a]	0.10	1.9484
		2006	0.15		0.11		0.13		0.9989	0.14	
	Land	1996	0.40	4.8362[a]	0.29	5.3999[a]	0.45	3.8568[a]	1.8753	0.38	7.9328[a]
		2006	0.59		0.66		0.73		4.5440[a]	0.63	
III	Physical capital asset	1996	0.21	4.7781[a]	0.24	2.8032[a]	0.21	2.9620[a]	0.5638	0.22	6.2813[a]
		2006	0.31		0.30		0.36		1.2453	0.31	
	Irrigation infrastructure	1996	0.21	4.1409[a]	0.25	0.0000	0.16	2.0545[a]	0.3014	0.21	4.1751[a]
		2006	0.44		0.22		0.30		2.2063[a]	0.36	
	Road infrastructure	1996	0.46	0.1648	0.62	-0.9978	0.61	1.5844	2.6033[a]	0.53	0.2063
		2006	0.44		0.56		0.62		4.4507[a]	0.50	
	Farm input	1996	0.36	0.3777	0.33	1.2386	0.35	1.0305	1.7519	0.35	1.4885
		2006	0.36		0.35		0.37		0.6729	0.36	
	Processing	1996	0.01	1.0000	0.00	1.9163	0.01	0.9282	0.5398	0.01	2.2008[a]
		2006	0.03		0.05		0.05		0.4721	0.04	

IV	Financial capital asset	1996	0.30	2.5709[a]	0.30	1.7747	0.31	−2.2054[a]	1.1699	0.30	1.8408
		2006	0.31		0.31		0.29		3.2637[a]	0.31	
	Cash income	1996	0.89	1.8763	0.89	1.8866	0.90	−0.6831	0.1993	0.89	2.1170[a]
		2006	0.90		0.91		0.89		1.7525	0.90	
	Credit	1996	0.01	−1.9504	0.01	−0.7641	0.04	−1.9122	1.8233	0.02	−2.7136[a]
		2006	0.00		0.00		0.00		1.3727	0.00	
	Subsidy-tax	1996	0.00	2.7774[a]	0.00	2.9523[a]	0.00	−0.7381	1.0107	0.00	2.9008[a]
		2006	0.03		0.02		−0.02		2.1784[a]	0.02	
V	Social capital asset	1996	0.41	−4.8943[a]	0.49	0.0444	0.56	3.5234[a]	8.1305[a]	0.45	−1.7467
		2006	0.34		0.49		0.69		149.4090[a]	0.43	
	Equity	1996	0.39	−4.7945[a]	0.53	1.1422	0.75	3.3969[a]	15.5588[a]	0.48	−0.8981
		2006	0.26		0.59		0.98		494.4920[a]	0.46	
	Institutional participation	1996	0.43	−1.7922	0.44	−2.4279[a]	0.37	0.7199	1.2204	0.43	−2.2530[a]
		2006	0.41		0.39		0.39		0.2564	0.40	

[a] *Significant at 95% confidence level.*

I. LIVELIHOOD DEPENDENCE, RIGHTS AND ACCESS TO NATURAL RESOURCES

period. This increase can be explained by the higher fertility rate of low- and middle-income households 10 years ago. The children that were born then have now reached working age, and this increase in the household workforce accounts mainly for the increased access to labor of low- and middle-income groups. Access to education has increased significantly for poor households, almost doubling their opportunities. The main reason for this increase is the provision of scholarships and allowances by the GoI for children from low-income groups after the economic crisis. Overall access to human capital assets has increased slightly with a greater increase for families with low and middle income.

Increase in access to natural capital assets was significant, and was also distributed equally among different income groups. However, access to the building variables of this asset and the change of these assets varies. Access to forest resources has increased insignificantly and poor households have had more access to forest resources in 2006 than in 1996. Moreover, access to land has increased substantially for all income groups; the high-income group has, however, experienced the highest rate of increase. There was uncertainty concerning the rules of access due to the restoration of *nagari* rule, the issuing of a new forest law, and weaker authority of local leaders on the *nagari*'s forestland. It is likely that this situation has caused local people to extract more timber and NTFPs from small forest plots in the study site than before. The access of poor households to forest resources is therefore uncertain in the long run. In a situation of competition for land, rich families are likely to obtain better security of land ownership and better quality of land.

Access to physical capital assets grew substantially during 1996 and 2006, but there was no significant difference in the overall index among different income groups. However, when examining building variables indexes there were considerable differences between income groups in 2006, especially with respect to access to road and irrigation infrastructures. Access to irrigation infrastructure has increased dramatically, and poor households' access has virtually jumped. The reason for this trend is the expansion of irrigation infrastructure to upland areas to maintain rice self-sufficiency. The GoI with financial support from the Asian Development Bank (ADB) has launched some irrigation projects since 2001 to maintain irrigation canals and to enlarge irrigation coverage. This is also true for the current site. Road infrastructure, however, has deteriorated due to lack of attention. Road facilities that were constructed two decades ago have not been maintained well, which led to a reduction in access of upland inhabitants to road infrastructure in 2006. Low- and middle-income households were affected to a larger extent by poor infrastructure than households from the high-income group. Access to farm input increased slightly and was distributed equally among different income groups. Contrary to our initial prediction, fertilizer usage increased despite a price increase. However, access to processing facilities has increased substantially with all income groups participating in this increase. This is mainly related to commercial rice production. Privatization of the rice market has encouraged rural rich households to become involved in rice marketing as well as to invest in small-scale rice milling. The number of rice milling unit (RMU) increased sharply in the study site from 11 units in 1996 to 24 units in 2006, particularly in *Nagari* Dilam and Koto Laweh. As a result of the increased availability of RMU, farmers are encouraged to sell processed rice.

With respect to the financial capital asset, overall, the access both increased and decreased in 2006, depending on income group in 2006. While the access to this asset by the low-income group has increased substantially, access by families belonging to the high-income group has been reduced. Poor households had more opportunities to earn cash income from remittances as the economic crisis forced them to find alternative sources of income, including working in urban areas. In addition, increasing access of low-income families to NTFPs is also the reason why they got higher access to cash income, as most of NTFPs were sold commercially. The subsidies, which are provided by the GoI to poor families, have also increased the access of poor families' to cash income. Furthermore, cash subsidies for the most vulnerable households has caused the subsidy-tax index of low- and middle-income groups to grow substantially while the index of rich families has been reduced. However, all income groups had less access to credit in 2006 as compared to 1996. The GoI policy to liberalize rural financial markets has increased the interest rate. At the same time, formal credit institutions imposed strict collateral requirements that limited the access of upland inhabitants to credit from formal lending institutions. That poor households' access to financial capital assets has increased while rich families' access has decreased is mainly due to structural adjustments of Indonesia's macroeconomy.

Access to social capital assets has been reduced slightly, and the differences in access to this asset are significant among income groups. Low-income families' access has been reduced substantially, and access of rich families has increased. Inequity has increased among the upland inhabitants due to rising incomes of rich families and declining incomes of poor families as a result of the economic crisis. Rich families obtain benefits from rural economic liberalization as they invest in rural small agricultural processing industries, such as rice milling, and in agricultural input and output trading. Furthermore, lack understanding of low- and middle-income households with respect to the process of reestablishing *nagari* institutions, including forest management and water user associations (WUA), has been the cause of limited participation. The institutional participation index has therefore been reduced significantly.

3.6.2 Changes in Livelihood Strategy

An analysis of livelihood strategies revealed three major strategy elements: migration, job diversification, and agricultural intensification. Some households combined two or three of these elements. People from different income groups differ with respect to which element they prefer. Low-income households are likely to migrate more, extract more forest resources, work in unskilled low-paying jobs, and earn more income from remittances. Households of middle and high income tend to improve their livelihood security by occupational diversification toward more nonagricultural and off-farm activities. Therefore, there is a decreasing trend for all household categories of deriving incomes from agriculture during the last 10 years (Fig. 3.5). In addition, there is a trend toward strengthening social ties within a lineage by organizing internal microsaving and microlending to cope with adversities as well as with difficulties to secure credit from formal lending institutions.

Out-migration is a strategy practiced by some local people in response to the insecurity of agricultural activities. These are mostly farmers who have left the *nagari* to earn additional income, particularly by working in urban areas. Fig. 3.6 presents the percentage of households with migrants, who are sending remittances home regularly. These remittances comprise nearly 20% of the income of households with migrants. Low-income families have larger percentages (25%) of their members working outside than families from middle- and high-income groups (17% and 8%, respectively). There are two main factors that force the

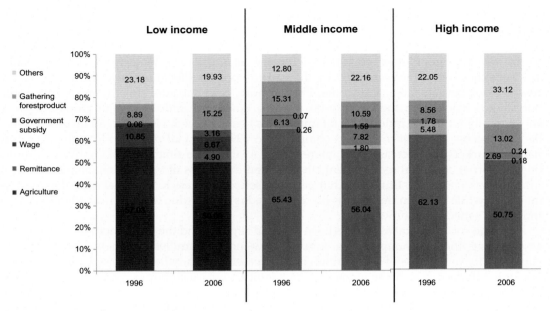

FIG. 3.5   Sources of household income.

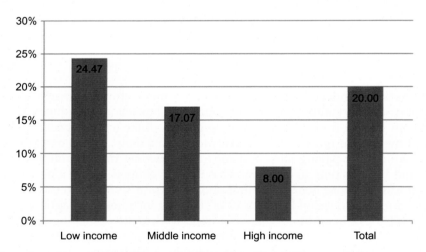

FIG. 3.6   Percentage of households with migrants who send remittances home regularly.

people with low income to find a job outside. First, young people from the low-income group have little opportunity for higher education, and are generally not attracted to working in the agricultural sector. Second, employment growth within the *nagari* is not sufficient to absorb the growing labor force, which comes mostly from low-income households. Despite the fact that the low-income group has more access to land, the marginal growth in agricultural jobs is sufficient only to provide labor for the adult household members. In addition, migration is also facilitated by a good network of transportation.

Diversification is another strategy chosen by marginal families in three *nagari*s of the research site. Although diversification has been practiced before, recent economic pressures have forced people to intensify this strategy. People diversify their livelihoods by raising cattle, by working as agricultural wage laborers, and through nonagricultural jobs. Fig. 3.6 shows the different sources of household income in 1996 and 2006 among the different groups. Agriculture is still the most important source of income, but has reduced its share over time in all income groups.

Low-income households had a greater diversity of income sources in 2006 as compared to 1996. The contribution of wage labor has diminished because some family members have migrated out, and the income has been compensated for by forest resources extraction. The contribution of this activity to the income of poor households increased from 8.9% in 1996 to 15.3% in 2006. Poor households have also received subsidies from government programs to compensate for income reduction due to structural adjustment policy. People from the middle-income group have increased their earnings either by finding new jobs within the *nagari* or through other activities such as trading, opening agricultural-processing small industries, and opening local petty shops (*warung/dagang keliling*). People from the high-income group have received less remittance because some of the family members who were working in urban areas had to come back due to the economic crisis, and are now engaged in local trading, in practicing intensive agriculture, and in raising cattle, which are comparatively less-remunerative activities.

Although the contribution of agriculture to household income has remained stagnant, there is a significant change in agricultural practices within the research site. Agricultural practices of middle- and high-income families have become more intensive. Use of high doses of agrochemicals in cultivating chili, tomato, carrot, and other crops has become common practice. Intensification has provided higher employment opportunities. However, because of intricate social relationships, new opportunities are made available to close family members only, and there is little chance for household members from the low-income group to share in this trend. This is the reason why households from the middle- and high-income groups obtained higher income from wage labor in 2006 as compared to 1996. Members of the high-income group are also investing in cattle raising, which provides them with the opportunity to take more grass from *nagari* land or from lineage land.

Finally, some lineages strengthen their social ties by enhancing the role of the lineage as a kinship institution not only for social but also for economic purposes. They have organized internal microsaving and microlending schemes in response to difficulties in getting credit from formal institutions. *Kaum Melayu* in *Nagari* Dilam, for example, manages savings and credits for its members in particular to cover those household expenditures that exceed regular income (eg, medical treatment), and elaborate traditional ceremonies such as weddings, an so on.

3.7 LIVELIHOOD AND IWM SUSTAINABILITY

Upland people in the Lembang subwatershed have been able to maintain and enhance their livelihood capabilities and their access to some assets. Changes of context have provided them with opportunities to have more access to natural and physical assets. Access to social capital, however, was reduced during the same period. Less access to social capital for the poor means that the new sociopolitical framework has not been able to encourage poor households to participate in the new institutional arrangements. Recent changes of context have, however, reduced income equity.

Livelihood sustainability is a complex phenomenon with several factors intervening and interacting with each other as becomes evident from the findings of this study. While access to some assets has improved, others have deteriorated. Livelihood sustainability is also an indicator of performance of watershed management according to Campbell et al. (2001) and Shivakoti and Shrestha (2005a).

Table 3.5 recapitulates the indication of livelihood sustainability. Changes in NRM context, change in access to capital assets. and change in livelihood strategies during the last 10 years at the study site shifted the livelihood sustainability of the upland inhabitant. Achieving environmental sustainability is an especially big challenge in the face of high agrochemical input and intensive soil tillage practices. The greatest environmental challenges in the Lembang subwatershed are loss of forest cover, decreasing water availability, and increasing soil erosion.

Access to those capital assets that are related to the economic aspect of sustainable livelihood has improved, while access to others has deteriorated. Increasing access to land, labor, and irrigation infrastructure are among the improvements of economic sustainability, by which the low-income group has improved livelihoods. Agricultural market liberalization, on the other hand, has hurt low-income groups as prices for basic goods have gone up and as the availability of low interest credit rates from formal lending institutions has decreased. Government efforts to provide subsidies must be viewed as only a temporary measure. Therefore, economic sustainability can only be maintained when the negative effects of agricultural market liberalization are brought under control.

TABLE 3.5   Livelihood Sustainability Indication in Upland of Lembang Subwatershed of West Sumatra

No	Livelihood Sustainability Aspects	Indication	The Situation in Upland of Lembang Subwatershed During Last 10 years
1	Environmental	Conserving or enhancing the productivity of life-supporting natural resources	• Degraded land increased • Forest cover decreased dramatically • Water erosion level increased • Intensive agricultural practices by using high agrochemical input and continuation of the depletion of natural resources
2	Economic	Maintaining the given level of expenditure	• Market liberalization hurts low-income groups. Government provides cash and noncash subsidies. However, the sustainability of such subsidies is under question • Less access to formal credit • Manpower of low-income group increased, while that of middle- and high-income groups decreased • Diversification of income sources has increased capacity of livelihood to cope with shock • Poor-household income has increased slightly

Continued

TABLE 3.5 Livelihood Sustainability Indication in Upland of Lembang Subwatershed of West Sumatra—cont'd

No	Livelihood Sustainability Aspects	Indication	The Situation in Upland of Lembang Subwatershed During Last 10 years
3	Social	Minimizing social exclusion and maximizing social equity	• Increasing income inequity among community groups • Increasing access to irrigation infrastructure for low-income household • Low- and middle-income groups have less access to road infrastructure, while high-income group gained more access • Access to education increased substantially, in particular, for people from low-income group • Access to land increased for the people from low-income families • Low-income groups gained more access to forest resources (*nagaris* common)
4	Institutional	Capacity of prevailing structures and process to continue	• Decentralization and return to *nagari* is a promising institutional model for natural resource management, but it is still in the beginning phase. Some conflicts occur both between local customary rule and national laws and between neighboring *nagari*s • New forestry and water resources laws provide more opportunity for people at local level to actively participate in management of these resources • Local institutions for forest and water management and local leaders' authorities over land and forest resources management are still weak due to change of land utilization rights

Increasing access of members of the low-income group to irrigation infrastructure, to forest resources, to land, and to education opportunities indicates a reduction in social exclusion. However, increasing income inequity between high-income groups indicates that middle- and high-income groups can still capture more of the benefits that result from the changing NRM context. Therefore, social sustainability has not yet been fully attained.

Institutional sustainability in the research site is also jeopardized. Local institutions for forest and water resources management are still weak. Conflicts over resource utilization still occur because detailed regulations have not yet been established. New national laws on decentralization and the management of forest and water resources as well as the restoration of *nagari* administration are in need of detailed guidance on institutional mediation processes. Responses from the local government as well as from the *nagari*s, however, are slower than the pace of economic development and the effects of economic liberalization. Newly reestablished *nagari*s and some WUAs are not yet powerful enough to handle conflicts over natural resource utilization. For IWM institutional sustainability, a stronger effort to strengthen local institutions is urgently needed.

3.8 CONCLUSION AND POLICY IMPLICATIONS

The findings from this research can be helpful both in explaining the effects of policy change on household livelihoods as well as in providing guidelines for sustainable watershed management. The paper started out by discussing policy change at national and local levels with respect to NRM for the livelihoods of the local population. Change in access to capital assets of upland households of the Lembang subwatershed has created economic pressure. Moreover, in response to shocks and cash needs, people exploit natural resources more intensively. This trend is further exacerbated when user rules are unclear due to changes in national laws and in the implementation of new governance arrangements such as decentralization and restoration of *nagari* rule that led to an increase in access to natural capital. The government has provided subsidies to help poor households who face difficulties in gaining access to some capital assets. Their access to social capital asset has, however, been reduced due to an increase in inequity that was caused mainly by the high-income group getting more benefits from market liberalization, and by poor households participating less in the newly reestablished institutions.

Our analysis of changing livelihood strategies also shows that households, in response to these changes in the NRM context, are increasingly searching for nonagricultural sources of income, and are gaining more direct access to capital assets. Even though the contribution from agriculture to household income has become less, agricultural practices have direct influence on environmental sustainability through intensification of land-use practices and high agrochemical inputs. Forest resources extraction, migration, and outside employment are the major activities of people from the low-income group, while people from middle- and high-income groups tend to increase their livelihood security by intensification of agriculture and by investment in trading activities and small rural industry. These shifts are indirectly related to policy changes.

The findings from this study also provide guidelines for watershed management by providing tools to assess management performance on the basis of environmental, economic, social, and institutional sustainability indicators. Continuing degradation of natural resources due to high agrochemical input has negative effects on the watershed environment. Economic sustainability, on the other hand, is indicated by the ability of poor households to increase their income. Social aspects of watershed management exhibit both negative trends—increasing income inequity—as well as positive trends: decreasing social exclusion. Institutional weakness requires immediate action from decision makers and natural resources managers.

Therefore, the policy implication of this study is that to improve livelihood as well as watershed sustainability immediate action must be taken to strengthen local institutions, to conserve natural resources, and to promote environmentally sound agricultural practices within the watershed.

We strongly believe that the *nagari*s can achieve effective management, because *nagari*s are reestablished on the basis of local customary laws (*adat*). *Nagari*s are not introduced by outsiders with some simplification that was criticized by Li (2002). Although newly established *nagari*s are still plagued by weaknesses, they will be able to function as strong local institutions in NRM according to findings of observers of this process of reestablishment (von Benda-Beckmann and von Benda-Beckmann, 2001; Nurdin, 2007) and supported by our own research findings. Current trends in Indonesia are promising with respect to providing an enabling environment for CBNRM (Armitage, 2005), also in West Sumatra. To develop strong and dynamic local institutions takes time and effort (Lam, 2001). The central government has to provide clear and certain laws that can guide people at the local level to improve their institutional capabilities. Recent laws on forest and water resources as well as the regional government acknowledge sufficiently the main role of customary laws. This task could be executed by local government, nongovernmental organizations (NGOs), universities, or collaborative actions among these organizations. Financial support from international donor agencies is also needed.

Along with strengthening of local institutions, conserving natural resources, especially forest resources, can be done in a participative way within the decentralization framework. For this purpose, the local government provides guidance and support for local institutions to formulate and implement protection and conservation rules. *Nagari*s should be empowered in two ways. First, they should be given a clear role in forest and water management. This role needs to be laid down and explained clearly in local government regulations on water and forest. Second, their management capabilities must be improved with respect to both human resources capabilities and equipment. To sum it all up, natural resources conservation will be more effective when the authority over it has been handed over to *nagari*s and other local institutions, after they have been empowered.

Promoting environmentally sound agricultural practices is also an urgent requirement. High agrochemical input has polluted water and soil in the subwatershed and affected lowlanders' social and economic activities. Intensive soil tillage on steep slopes is another threat. To promote organic farming could be a promising solution. It is extremely urgent that the GoI carries out action to promote organic farming with support from NGOs and local universities.

However, this research has not yet clarified which one of the above contexts has the strongest and most direct impact on livelihood change and livelihood. Further research is still needed to answer this question. Probably, a different methodology is also needed to increase the number of building variables of access to capital assets to link quantitatively the change in NRM context with the change in livelihood and livelihood sustainability. The new method could contribute significantly to livelihood change studies in the future.

Acknowledgments

It is duly acknowledged that this study was funded by a grant of the Ford Foundation-Jakarta Office to Andalas University and the Asian Institute of Technology. We are grateful to local residents of Lembang subwatershed for their participation in interviews and surveys. An earlier version of this paper was published in Mahdi, Shivakoti, G.P., Schmidt-Vogt, D., 2009. Livelihood change and livelihood sustainability in the uplands of Lembang subwatershed, West Sumatra, Indonesia, in a changing natural resource management context. Environ. Manage. 43 (1), 141–165, which is duly acknowledged.

References

Armitage, D., 2005. Adaptive capacity and community-based natural resource management. Environ. Manage. 35, 703–715.

Armitage, D., Marschke, M., Plummer, R., 2008. Adaptive co-management and the paradox of learning. Glob. Environ. Change 18, 86–98.

Arnold, J.E.M., 1998. Forestry and sustainable rural livelihoods. In: Carney, D. (Ed.), Sustainable Rural Livelihood What Contribution Can We Make. Department for International Development (DFID), London, pp. 155–166.

Ashley, C., Carney, D., 1999. Sustainable Livelihoods: Lessons From Early Experience. Department for International Development (DFID), London. 55 pp.

Bebbington, A., 1999. Capitals and capabilities: a framework for analyzing peasant viability, rural livelihoods and poverty. World Dev. 27, 2021–2044.

Campbell, B., Sayer, J.A., Frost, P., Vermeulen, S., Porez, M.R., Cunningham, A., Prabhu, R., 2001. Assessing the performance of natural resource systems. Cons. Ecol. 5, 22–45.

Carney, D., Drinkwater, M., Rusinow, T., Neefjes, K., Wanmali, S., Singh, N., 1999. Livelihoods Approaches Compared: A Brief Comparison of the Livelihoods Approaches of the UK Department for International Development (DFID), CARE, Oxfam and the United Nations Development Programme (UNDP) (DFID Working Paper). Department for International Development, London.

Chambers, R., 1995. Poverty and livelihoods: whose reality counts? Environ. Urban. 7, 173–204.

Chambers, R., Conway, G.R., 1992. Sustainable Rural Livelihoods: Practical Concepts for 21st Century. (IDS Discussion Paper 296). Department for International Development, London.

Dewi, S., Belcher, B., Puntodewo, A., 2005. Village economic opportunity, forest dependence, and rural livelihoods in East Kalimantan, Indonesia. World Dev. 33, 1419–1434.

DFID (Department for International Development), 1999. Sustainable Livelihoods Guidance Sheets. Department for International Development, London. 92 pp.

Dupar, M., Badenoch, N., 2002. Environment, Livelihoods, and Local Institutions Decentralization in Mainland Southeast Asia. World Resources Institute, Washington, DC. 70 pp.

Feridhanusetyawan, T., Pangestu, M., 2003. Indonesian trade liberalisation: estimating the gains. Bull. Indones. Econ. Stud. 39, 51–74.

Gellert, P.K., 2005. The shifting natures of "development": growth, crisis, and recovery in Indonesia's forests. World Dev. 33, 1345–1364.

Haryati, Y., Aji, J.M.M., 2005. Indonesian rice supply performance in the trade liberalization era. In: Paper Presented at the Indonesia Rice Conference 2005, Tabanan Bali 12–14, Sep. 2005.

Irianto, G., Surmaini, E., Pasandaran, E., 2004. Dinamika iklim dan sumber daya air untuk budi daya padi. In: Kasryno, F., Pasandaran, E., Fagi, A.M. (Eds.), Ekonomi Padi dan Beras Indonesia. Badan Penelitian dan Pengembangan Pertanian, Departemen Pertanian Republik Indonesia, Jakarta, pp. 255–276. Ekonomi Perberasan Indonesia.

Istijono, B., 2006. Konservasi Daerah Aliran Sungai dan Pendapatan Petani: Studi Tentang Integrasi Pegelolaan Daerah Aliran Sungai. Studi Kasus DAS Sumani Kabupaten Solo/Kota Solok, Sumatera Barat. (Unpublished-Dissertation). Universitas Andalas, Padang.

Koeberlein, V.M., 2003. Living From Waste: Livelihood of the Actors Involved in Delhi's Recycling Economy. Verlag für Entwicklungspolitik, Saarbrücken.

Lam, W.F., 2001. Coping with change: a study of local irrigation institutions in Taiwan. World Dev. 29, 1569–1592.

Li, T.M., 2002. Engaging simplifications: community-based resource management, market processes and state agendas in upland Southeast Asia. World Dev. 30, 265–283.

Mitchell, B., 2002. Resource and Environmental Management. Prentice Hall, Harlow. 367 pp.

Mitchell, B., 2005. Integrated water resource management, institutional arrangements, and land-use planning. Environ. Plan. 37, 1335–1352.

MoF (Ministry of Forestry), 2003. Report to Stakeholders: Current Condition of Forestry Development. Ministry of Forestry (MoF), Jakarta.

Naim, M., 1984. Merantau: Pola Migrasi Suku Minangkabau (Merantau: Migration Pattern Among Minangkabau Ethnic). Gajah Mada University Press, Yogyakarta.

Nurdin, A., 2007. Resolusi Konflik Tanah Ulayat di Minangkabau, Sumatera Barat. Studi Kasus Tujuh Nagari Konflik di Sumatera Barat (Unpublished-Dissertation). Universitas Andalas, Padang.

O'Connor, C.M., 2004. Effects of central decisions on local livelihoods in Indonesia: potential synergies between the programs of transmigration and industrial forest conversion. Popul. Environ. 25, 319–333.

Resosudarmo, I.A.P., 2002. Closer to people and trees: will decentralization work for the people and the forests of Indonesia? In: Paper Presented at the World Resources Institute Conference on Decentralization and The Environment, Bellagio, Italy, 18–22, Feb. 2002.

Robinson, S., El-Said, M., San, N.N., 1998. Rice policy, trade, and exchange rate changes in Indonesia: a general equilibrium analysis. J. Asian Econ. 9, 393–423.

San, N.N., Löfgren, H., Robinson, S., 2000. Structural Adjustment, Agriculture, and Deforestation in the Sumatera Regional Economy (TMD Discussion Papers). International Food Policy Research Institute, Washington, DC.

Scoones, I., 1998. Sustainable Rural Livelihoods: A Framework for analysis (IDS Working Paper 72). Department for International Development, London.

Shivakoti, G., Shrestha, S., 2005a. Analysis of livelihood asset Pentagon to assess the performance of irrigation systems: part 1—analytical framework. Water Int. 30, 356–362.

Shivakoti, G., Shrestha, S., 2005b. Analysis of livelihood asset Pentagon to assess the performance of irrigation systems: part 2—application of analytical framework. Water Int. 30, 363–371.

Statistics of West Sumatra, 2005. Sumatera Barat Dalam Angka 2004 (West Sumatra in Figures 2004). Padang, Badan Pusat Statistik Sumatera Barat (BPS).

Torras, M., 1999. Inequality, resource depletion, and welfare accounting: applications to Indonesia and Costa Rica. World Dev. 27, 1191–1202.

Vedeld, P., Angelsen, A., Bojo, J., Sjaastad, E., Kobugabe Berg, G., 2007. Forest environmental incomes and the rural poor. Forest Policy Econ. 9, 869–879.

von Benda-Beckmann, F., von Benda-Beckmann, K., 2001. Recreating the Nagari: Decentralization in West Sumatra. (Working Papers No 31). Max Planck Institute for Social Anthropology, Halle.

WCED (World Commission on Environment and Development), 1987. Our Common Future. Oxford University Press, Oxford.

WRI (World Resources Institute), 2001. World Resources 2000–2001: the Fraying Web of Life People and Ecosystems. World Resources Institute (WRI), Washington, DC.

WRI (World Resources Institute), 2005. World Resources 2005: the Wealth of the Poor—Managing Ecosystems to Fight Poverty. World Resources Institute (WRI), Washington, DC.

Yonariza, Shivakoti, G.P., 2008. Decentralization policy and revitalization of local institutions for protected area co-management in West Sumatra, Indonesia. In: Webb, E.L., Shivakoti, G.P. (Eds.), Decentralization, Forests and Rural Communities: Policy Outcomes in South and Southeast Asia. Sage, New Delhi, pp. 128–149.

A Case Study of Livelihood Strategies of Fishermen in Nagari Sungai Pisang, West Sumatra, Indonesia

R. Deswandi[a]

National Project Manager at the United Nations Industrial Development Organization, Jakarta, Indonesia

4.1 INTRODUCTION

4.1.1 Background

Nature is seldom linear, but rather dynamic and unpredictable (Berkes et al., 2003). Therefore, humans have to retain adequate knowledge of adaptive capacity to secure their livelihoods against dynamic environmental changes and disturbances. Fishermen whose livelihoods rely on unpredictable fishery resources would experience a higher level of uncertainty compared to farmers whose livelihoods depend on modifiable agriculture activities. Farming may involve a wide range of management controls (land tillage, manure, and pest control) through which farmers may improve the quality of their land, select only highly productive crops, and so forth. Such intervention reduces the risk of loss and the level of uncertainty in farming. Schlager and Ostorm (1999) documented common pool resource dilemmas that might be experienced by fishermen.[1] Natural constraints, such as seasonal variability, might severely affect fishing activities. In the tropic region, during April to October (the summer monsoon), fishing might not always be feasible. During the summer monsoon, tropical storms may unpredictably strike at anytime. This situation was observed in Nagari Sungai Pisang, West Sumatra, where artisanal

---

[a] Former student at the Department of Integrated Natural Resource Management at Andalas University.

[1] The dilemmas are appropriation externalities, technological externalities, and assignment problems. Appropriation externalities occur as fishermen withdraw fish from a common stock without taking into account the effects of their harvesting upon each other. Technological externalities refer to the situation where fishermen physically interfere with each other during harvesting activity. The assignment problem is inherent with the fact that fish stock varies across fishing grounds, and tends to congregate in particular areas. The latter condition triggers competition.

fishermen were unable to go fishing during that time. The intensity of fishing activity in the area was reduced accordingly. In Nagari Sungai Pisang, this seasonal/climatic phenomenon was not the only constraint. Coral reef bleaching in their coastal water was concurrent with a decreasing stock of economically important fish species (ie, anchovy, or the locally called *bada karang*). Similar to many cases of artisanal capture fisheries in Indonesia, the size of the catch per fishing activity often can't compensate fishing efforts (Pet-Soede et al., 2001).

With these prevailing constraints, fishermen in Nagari Sungai Pisang experienced fishing uncertainty; that is, a situation where they experienced a severe variability of the amount of daily catch to the level where fulfilling their daily subsistence was difficult. How do the livelihoods of fishermen in the area survive in the face of such uncertainty? As an integrated part of their nature, the sustainability of natural resources would affect the sustainability of livelihoods of the community whose livelihoods depend on the natural resources system, and vice versa. This assumption emphasizes the connectedness between a social system and an ecological system, especially as the community's livelihood depends solely on extraction of natural resources (Adger, 2000), as do the community's of fishermen.

4.1.2 Problem Statement

An observation in Nagari Sungai Pisang revealed that a considerable number of local people still performed artisanal fishing as their main activity (as defined by the total amount of time they spent at it daily). This, arguably, indicated the ability to cope with the inherent uncertainty in fishing to sustain their livelihoods. How did they manage to sustain their livelihoods under such circumstances? What strategies and measures were taken to ensure the availability of food and cash for their families? These are major questions in this research. To answer these questions, this qualitative research set the following objectives: (1) to identify livelihood strategies developed and activities carried out by fishermen as a means to obtain food and/or cash, and (2) to identify livelihood assets that support the development of livelihood strategies.

4.2 RESEARCH METHODOLOGY

This qualitative study focused on the process through which people make their livings, experiences, and constructed their world rationale to themselves as manifested through their words and/or actions (Creswell, 2009; Afrizal, 2005; Stainback and Stainback, 1988). Primary data (transcripts of interviews, field notes, pictures, and sketches) were collected during fieldwork from May to August 2007.[2] Data[3] were classified as primary and secondary data.

---

[2] Observation and interaction with the local community, especially with fishermen, have been carried out and established since 2003 when the researcher actively participated as a volunteer in the Coral Reefs Rehabilitation and Management Program (COREMAP). In the program, fishermen and their families were the main targeted stakeholders. Interaction with local community in the *nagari* continued as the researcher was involved in three research studies in the coastal water in 2004 and 2006. During these studies, the researcher visited Nagari Sungai Pisang every week and spent 2–3 days. Intensive interaction with key informants was initiated during the preparation of fieldwork in February 2007. Through this history of interaction, the presence of the researcher in the field should have been nonobtrusive.

[3] Livelihood strategies are a means for outcomes to ensure livelihood security. This research limited data collection to relevant information on strategies and activities developed and carried out to obtain food and/or cash.

Primary data were collected through semistructured in-depth interviews and participant observation. Semistructured in-depth interviews were informed by interview guidance (a set of main starting questions). Participant observation was performed at a moderate level or type of participation (Stainback and Stainback, 1988). Two procedures of triangulation were employed, such as triangulation by data sources and methodological triangulation (Stainback and Stainback, 1988).

Individual fisherman is the unit of analysis of this research. To select informants out of 141 listed names,[4] a secondary database obtained from the Kelurahan office was first verified. Potential informants were shortlisted through set criteria and indicators of eligibility[5] (purposive selection) developed to ensure the validity of the research's results/ findings. Of 141 names, only 33 fishermen were eligible for this research. Taking into account the preference for individuals with respective social status and/or acknowledged fishing knowledge and experience, six key informants were selected. They were Bsw (62 years old), Is (49 years old), SPm (50 years old), Wn (50 years old), Al (59 years old), and Az (55 years old).[6] Nevertheless, data collection activity was extended to other fishermen, including customary figures or the *panghulu*[7] and the head of government representative in the *nagari*, to gain deeper information and wider perspective on research problem. To validate data, a number of interviews were carried out with several groups of men and women during a village meeting.

Nagari Sungai Pisang is administratively known as Kelurahan[8] Teluk Kabung Selatan. This area is located in the subdistrict Bungus Teluk Kabung, Padang Municipality (E 1000 20′ 15″ to 1000 24′ 15″ and S 10 5′ 0″ to 10 8′ 15″) with total area of 9.14 km² (BPS, 2004). Situated 10 km from the subdistrict and 31 km from the municipality, Nagari Sungai Pisang is the home of 1714 individuals, comprising 851 males and 863 females (Fig. 4.1).

Nagari Sungai Pisang covered areas of land and coastal water, including five small islands; namely, *Pasumpahan, Sirandah, Sikuwai, Sironjong*, and *Setan* islands. An approximate

[4] According to a report on a socioeconomic survey carried out by Yayasan Hayati Lestari (a local NGO in West Sumatra), there were 141 fishermen remain in Nagari Sungai Pisang as per March 2006.

[5] Criteria for the selection of informants were an individual fisherman who (1) has been fishing for (at least) 25 years at the time of field research (in 2007), (2) has to have his own fishing equipment that supports fishing activity, (3) has been married or serving as the main economic backbone of his family, and (4) has considerable fishing experience that is acknowledged by the locals. Those who have respective status within the local community (eg, in the customary structure/system) were preferred for informants. The indicators for eligible informants included that they spent more than 6 h/day fishing (except during the summer monsoon) and they operated during the time the field work was conducted (for an observable activity). The most recommended fishermen for their knowledge, experience, and understanding of the problems under study were the key informants of this research.

[6] Considering research etiquette and to protect the privacy of key informants, their names (and so the names of informants) are written with their initials.

[7] *Panghulu* (Minangkabau language) or *Penghulu* (Bahasa Indonesia) stems from the "*hulu*," which means the upstream area of rivers. It equals with the top of a structure, which implies superiority or authority (eg, to decide). Therefore, *penghulu* means persons (always adult males) who are authorized to make decisions for their community.

[8] *Kelurahan* is an administrative area under the subdistrict level. The official head of a *Kelurahan* is an officer called the *Lurah*.

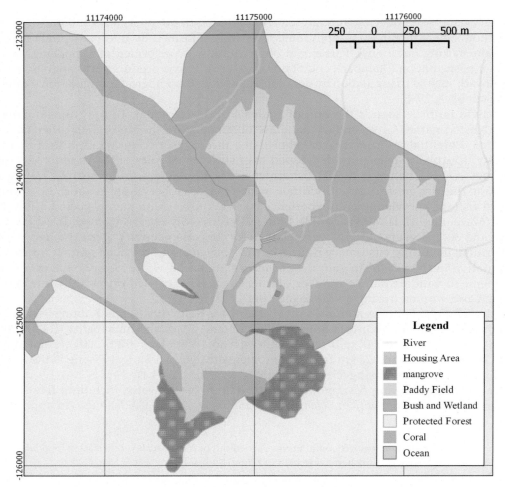

FIG. 4.1   Map of Nagari Sungai Pisang. *Modified from Land-use map of Sungai Pisang Village; Yayasan Hayati Lestari, 2006. Rencana Strategis Pengelolaan Kawasan Pesisir dan Laut Sungai Pisang, Yayasan Hayati Lestari and Dinas Kelautan dan Perikanan Kota Padang, Padang (Courtessy of Yulian Anggriawan, 2016).*

20 ha of mangrove forest laid in the bay area protecting some part of adjacent shoreline from prevailing coastal abrasion. Coral reef surrounded small islands and laid in the shore line. All reefs were subjected to bleaching and destructive blast fishing and cyanide fishing. Massive coral bleaching was caused by a global phenomenon El Niño Southern Oscillation (ENSO) occurring in 1997. ENSO also triggered a massive forest fire in Sumatra. For many days, smoke generated by the forest fire blocked the sunshine and was predicted to also be responsible for the death of coral reefs on the west coast of Sumatera. Reefs were also excavated both for ornamental reef collection and as a means of house protection from coastal abrasion.

I. LIVELIHOOD DEPENDENCE, RIGHTS AND ACCESS TO NATURAL RESOURCES

4.3 RESULTS AND DISCUSSIONS

Pragmatism in utilizing and maximizing whatever available resources and skills they have and the strong will to create or develop opportunities for sustaining their livelihoods are well-represented by traditional proverbs commonly cited by fishermen; *indak kayu janjang dikapiang* (if no fire woods are available, then use wooden stair for fuel), and *tak ado rotan aka pun jadi* (if no rattan is available, then three roots will do). These proverbs convey the idea of coping with unpredicted and unwanted situations in which the main and only objective will be to secure their livelihoods; that is, by providing food and/or cash, by any means possible, especially if the main livelihood activity failed, either because of environmental constraints or internal limitations (eg, health issues or lack of useful resources).

Memories of past events contributed to motivating fishermen to develop and employ strategies to a more sustainable livelihood. A decrease in the stock of economically important target fish after 1986 was the most shocking event for the livelihood of fishermen, where almost all fishermen lose their source of main cash income. Uncertainty in the amount of daily catch, whereby their livelihoods were constantly compromised, had been a common and prominent daily case experienced by fishermen in Nagari Sungai Pisang. To describe this situation, fishermen in the area often referred to the term *rasaki harimau* or the fortune of the tiger to perceive unpredictable results from fishing activity. In such a situation, fishermen typically said *"kadang indak ado pitih pambali bareh, tapi ado pitih pambali ameh,"* which means in a certain situation they failed to catch a sufficient amount of fish to fulfill their family's subsistence, but in other circumstances the catch was abundant that they could even invest in gold. *Rasaki harimau* depicted the idea of fatalism, where the result of fishing efforts is totally predetermined by God as the superior entity. By perceiving such an idea, fishermen could relieve the stressful situation derived from an uncertain fishing catch.[9] Perceiving the phenomenon of *rasaki harimau* to explain certain situations beyond their control has seemed to help them face stressful situations. Fishermen in Nagari Sungai Pisang agreed that *rasaki harimau* is one common phenomenon they should not be frustrated about.[10] In a social organization there is a range of more or less satisfactory ways to cope with aspects of uncertainty in daily life situations (Von Benda-Beckmann et al., 1994). Religion or belief has been a dominant factor to cope with environmental uncertainty performed by fishermen in Nagari Sungai Pisang, where almost all fishermen embraced and believed in Islam.

Livelihood uncertainty, as reflected in the term *rasaki harimau*, has become the main factor motivating the development of strategies to sustain livelihood. Accepting the phenomenon of *rasaki*

---

[9] Based on observations carried out every morning (the time fishermen returned from fishing), laughing and joking were a more common response to unfortunately small catch rather than sadness or disappointment.

[10] Individual or group perception on uncertainty and the extent of willingness to live with it would depend on, among others, psychological constitution. Clinical psychology research carried out in Aceh after a tsunami disaster in December 2005 explains this phenomenon. He found that when affected people consider disaster and the loss they experience as a situation beyond their control (in his case, as predetermined by God), the level of stress is lower compared to people who accept the situation the opposite way (Lengkong, 2007). This research conclusion differs with mainstream clinical psychology theory on locus of control. Such an "anomaly," he concluded, is explained by the sociocultural contexts of people in Aceh as devoted Moslems. Values and directives in Islam influenced the way they perceived daily events, including disaster.

harimau as part of daily life, fishermen were aware of the potential threats to their livelihood from, for example, food shortage. Informants revealed the time when they could not go fishing for many days and when they could not get a sufficient amount of catch. As fishing uncertainty has occurred in the past, fishermen realized that this would occur in the present and become worse in the future. Fishermen accepted that fishing uncertainty is inherent with this activity as the nature of resources they are dependent on and the livelihood activity they are committed to.

4.3.1 Activities for Food and Cash

To be able to provide daily needs for their family, fishermen employed more than one activity or efforts aimed at obtaining food and cash (livelihood activities). Moran (1982), Low et al. (2003), Davidson-Hunt and Berkes, 2003, and Folke et al. (2003) documents strategies of human adaptation to environmental fluctuations and resources uncertainties to maintain their family's well-beingness by fulfilling basis subsistence needs. The documentation made by those researchers was used as the basis for classification of livelihood strategies developed and performed by fishermen in Nagari Sungai Pisang.

4.3.1.1 Relying on the Kindness of Nature

This is the very basic strategy employed by hunters and gatherers and pastoralists, as it entails the least technological requirements to harvest whatever valuable resources are in the surrounding environment. Their settlement patterns usually reflect people decisions about how to make their living (Moran, 1982). Through this strategy, fishermen simply extract valuable resources directly from their surrounding environment. Historical information revealed that the first group of ancestors that decided to inhabit Nagari Sungai Pisang acknowledged abundant natural resources and rich landscape of the area as the main consideration to establish a settlement area. The availability of extractable resources was critical at the initial step for the area's development. A continuous supply of fresh water, abundant stock of fish and trees, and considerably fertile land in the area were important.

Fishing, which is simply harvesting the fish from the coastal water, was still the main livelihood activity performed by the fishermen in terms of time allocated to the activity. Every day, year round, fishermen spent 7–12 h fishing (depending on fishing equipment and weather conditions). Starting the activity in the afternoon, they finished fishing by dawn and returned to the coast.

Extracting resources from nature is a strategy practiced by communities with a strong connection to the natural resource system (Moran, 1982). Another activity of this strategy in the area was collecting firewood cooking fuel. Collecting abundant firewood reduced people's daily expenditures rather than using expensive kerosene to cook despite its convenience.[11]

Freshwater supply was critical for livelihood security. Most households received constant supplies year round through pipes from an abundant source situated in the upstream area of the *nagari*. However, a small number of households still relied on a waterwheel and water supplies installed in the public facility. To get their house piped with a water supply, every household must pay a monthly maintenance fee and, most importantly, pay for their own pipes.

---

[11] It was observed that, in the backyard of almost every house, there were piles of firewood that could serve for a half month.

4.3.1.2 *Alternative Productive Activities*

In areas where availability of and accessibility to resources are unpredictable, a generalist strategy is pursued (Davidson-Hunt and Berkes, 2003). A strategy through which fishermen diversified their activities for food and cash was employed by fishermen in Nagari Sungai Pisang. Fishermen in the *nagari* employed more than one activity to compensate for uncertain outcomes from fishing. Paddy cultivation, providing seasonal services for tourist visitors, making trawl (based on demand), and working as a shipbuilder were among alternative productive activities for food and/or cash.

Paddy cultivation was the most important alternative activity. It was capable of providing rice, a staple food for the local people in the *nagari*. Only a few fishermen did not employ this activity for the following reasons; either they did not have land, sufficient power, or time. A phenomenon where fishermen performed farming was also common in the *nagari*.

After 1986, many fishermen totally stopped fishing and then shifted to land-based activity, mostly paddy cultivation. These former fishermen confirmed that such shifting from fishing to paddy cultivation is still occurring in the *nagari*. Secondary data confirmed such a tendency. Total areas of paddy cultivation increased to more than 20% only within 2 years, from 59 to 73 ha (Yayasan Hayati Lestari, 2006; BPS, 2004).

To carry out paddy cultivation, some fishermen preferred hiring local laborers for almost all of the required cultivation activities, while others preferred doing it with the help of their family members and/or through collaboration with other fishermen/farmers. In the last case, they established informal groups (usually consisting of 10 people) that provided helped to members. The group would work at the paddy field of all members according to an agreed schedule. Hence, they did not have to pay anyone for paddy cultivation. They called this informal group a *balambiak hari*.

One consequence of such collaborative teamwork was all members had to allocate and adjust their daily activities to meet the agreed upon group's working schedules. It was imperative to maintain good performance and commitment within the group. Unacceptable performance and commitment would cause exclusion from the group. Having a bad reputation would end with refusal from any group, and therefore losing opportunities to get necessary help. Some fishermen preferred doing all activities in paddy cultivation on their own, because they were afraid of the inability to meet the group's schedule. Good reputation and acceptance in particular social networks have appeared to be important issues in the *nagari*.

Cash money for supporting paddy cultivation was provided, mostly, from fishing. *"Pitih dari lawuik ditanam ka darek"* (the cash from the sea was invested in the land). To engage in such investment and fulfillment of their family's need, fishermen agreed on important roles for housewives in spending and, at the same time, saving the money. There was another proverb to describe the importance of cash management within every family. The proverb says *kalau ado indak dimakan, kalau indak ado baru dimakan* (if you have something don't eat it, if there isn't anything then you may eat it).[12]

Nagari Sungai Pisang was also among tourism destinations. One of its small islands, Sikuwai Island, was a famous favorite leisure destination in West Sumatra. Local and

---

[12] The proverb refers to a practice of saving, where in a situation in which a family has cash money, it is advised not to spend all of it, but to save some portion of the money to anticipate unexpected future events.

international tourists visited the *nagari* both in transit to Sikuwai or as their main fishing or diving destination, including for research. Interaction between visitors and fishermen creates business opportunities where fishermen (and other locals) provided room accommodations, food, rentable boats, or as a local guide. However, such opportunities were often seasonal and therefore not fully reliable as a constant source of cash income.

Relevant practical skills help in creating side jobs to be performed after fishing. Many fishermen were capable of making or fixing fishing nets.[13] However, only a few of them consistently took this as a source of cash income. Demand was relatively low, and came mainly from outsiders. One fishing net was sold for IDR 1–2 million (approximately US $100–200) depending on its difficulties (mesh size) and the length and depth. One handmade fishing net might be produced within 6–12 months, because production took place only in spare time. Another practical skill was shipbuilding, which was a rather rare skill among fishermen and the locals. This skill was only shared within the family or with relatives.

There was a new option of alternative livelihood activity in the *nagari*. Few fishermen started establishing cutch plantation at the hillside (which was known as *ulayat nagari*) after obtaining a license from the *panghulu*. This activity and the necessary skills were bought to them by their relatives from another area known for its cutch plantation (ie, Nagari Siguntur, Subdistrict Pesisir Selatan). They saw a promising future in this activity after observing the better livelihood status of their relatives.

4.3.1.3 Forecasting for Unexpected Future Events

Particular activities were meant to anticipate unexpected needs; for example, medical treatments, school fees for their children, and other unforeseen family expenditures. Activities in this strategy served as insurance, such as saving some portion of their catch for daily consumption, preservation of fish, saving extra cash, participating in cash-saving groups (*julo-julo*), and raising livestock.

Fishermen always preferred saving a small portion of their catches for their families, mainly the less economically lucrative catch. Sometime they had to take a few economically valuable target fishes for the family's consumption to fulfill their obligation to their families (providing food). They have found out that it was more beneficial to save some fish for their families rather than selling all the fish for cash, and then buying other sources of protein that were often more expensive. When they have a larger side catch (less economically valuable fish), they traditionally preserved the fish for future consumption of their families, particularly for a time when it was impossible to go fishing.

Raising livestock (chicken, cows, and goats) was practiced particularly to serve as livelihood insurance. In a situation where fishermen urgently needed a substantial amount of cash, livestock might be sold or used as collateral to get credit or cash from others. To fishermen, this practice was similar to saving money. There was sufficient grassland area in the *nagari* to support this activity.

Accepting uncertainty as part of their livelihood, saving extra cash and/or ensuring the availability of valuable items to get cash money (eg, gold and livestock) have become

---

[13] Based on observation, most fishermen spent their spare time by talking in-group on the coastline near their boats while fixing their damaged fishing net. Sustaining damage of their nets was a common case. This repair skill was very important because their nets were their main asset for fishing.

important measures. Saving seems to be a common activity performed by communities facing livelihood uncertainty; however, it might take various forms. Such a practice stemmed from fishermen's accumulated day-to-day experiences that fishing might not always be reliable for getting food and/or cash money for their families. Through saving, they meant to enhance livelihood security against unexpected events in the future.

Joining a cash-saving group (or *arisan*) could be an alternative means to save money. Fishermen formed a group consisting of 10–15 individuals (not all of them were fishermen) where they periodically (daily or weekly) contributed an agreed upon certain amount of money to the group's treasury. Every month, the total amount of money that has been collected will be awarded to one selected member according to an agreed upon rule. A complete period of cash saving ended when all members had been awarded money. The group then decided whether or not to start another period, maybe with new members, new amounts of contribution, and/or a new period of awarding. Relying on this cash-saving group to get cash money, the longest awarding period was only 1 month. There also were groups practicing a shorter period of cash saving (once every week). This kind of group was locally called a *julo-julo*.[14] A considerable number of *julo-julo* groups existed in Nagari Sungai Pisang. However, a fisherman only chose to join the one he fully trusted the most. Before making the decision to join a particular group, the three most important criteria to fishermen were that the treasury of the group be known as a trusted person, the group's members also be trustable persons, and the awarding period and/or the whole *julo-julo* period be relatively short (the latter means not too many members in one group). Similarly, a fisherman will not be accepted by any group if he has a bad reputation for not contributing to his former groups. As the deciding person, a group treasury would select a group member by relying on one source of news: gossip. To join a cash-saving group to provide an alternative measure for cash saving, fishermen needed to consistently perform or act in accordance with common social norms; in this case, honesty.

4.3.1.4 *Maintaining or Enhancing Social Network*

An observation was made on the practice of catch sharing involving fishermen, the locals, and also outsiders when one fishing group happened to catch a huge school of fish very near to the shoreline. Realizing one group was struggling to fish out a significant amount of fish, other fishermen approached the group of fishermen, and so did the locals who happened to hear the news. When the fish were collected on the boat, all fishermen and the locals who surrounded the fortunate boat received three to five fish. This catch sharing was also observable every morning on the shoreline, when fishermen returned home. Despite the size of their catch, fishermen would share some portion of their catch to whomever asked for fish (mostly adult females). This, as fishermen named it, was *adaik pasia* (or the norm on the coastline). Referring to this norm, sharing some portions of their catch with family (for their subsistence) is an appropriate conduct. Hence, the opposite is an inappropriate one.

The act of catch sharing performed by fishermen was not entirely altruistic. By sharing their catch with other families, fishermen actively enhanced the possibility for having reciprocal

---

[14] As an illustration, one group agreed to collect IDR 10,000 per week, and awarded the total amount collected every month. Having eight members in the group, after 1 month, a total of IDR 320,000 was awarded to each member. It also means that the length of the cash-saving period of this group was 8 months.

support from those families they have helped. For instance, they would expect some sharing from other fishermen or the locals in a situation when they did not catch a sufficient amount of fish.

The practice of catch sharing is the act of extending a social relationship or network. The larger the network, the more possibility one has of getting support. On the contrary, those who did not practice catch sharing would reduce the chance of having support from others. Fishermen unwilling to share small portions of their catch would be labeled negative. At worst, those fishermen could be subjected to social exclusion and, consequently, they might not get any help whenever it was needed. Coping with livelihood security through a social network is also practiced by the fishing community in South Sulawesi, Indonesia. Fishermen and farmers organized themselves in a social organization called a patron-clients group (Deswandi, 2012; Glaser et al., 2010; Sallatang, 1982). Such a social organization in capture fisheries stems from the prevailing social and cultural contexts (Pelras, 2000, 2006).

Extending a social network through reciprocal generosity (catch sharing and cash-saving group) practiced by the fishermen in Nagari Sungai Pisang was also coping through adjustment of the social organization. Catch sharing in Nagari Sungai Pisang, arguably, has shifted from an altruistic conduct to an active livelihood strategy for extending the social relationship/network, and hence, social insurance through reciprocal generosity. This practice has been institutionalized as a norm. As a norm, it applied in the *nagari*, including to outsiders who happened to fish in its coastal water.

It was considered critical to extend this network of reciprocal generosity to the widest range of occupations; for example, to farmers and local merchants. Hence, fishermen could expect different kinds of help and/or assistance. While farmers might be expected to share their crops (rice), fishermen also relied on debt facilities from local merchants. In a situation where the results from fishing were extremely low or when tropical storms prevailed for days, many fishermen would have to rely solely on debts provided by local merchants whose kiosks could provide rice, sugar, and other basic needs. Sometimes, fishermen could also borrow cash money from them.

4.3.1.5 Risk Spreading Mechanisms and Productivity Enhancement

The first strategy performed by fishermen to respond to a decreasing stock of anchovy (the main target) in its coastal water was changing their fishing equipment and methods to fit catching other target fish. At the time of the research, one economically important target fish was mackerel (locally called *gambolo*). To catch mackerel, they fished at night and deployed more than one set of nets. To increase their chance to getting more fish, fishermen employed more than one type of fishing equipment to target many kinds of fish in different seasons. This particular fishing method required teamwork. Each member of the team contributed different input (ie, fuel, food, equipment, manpower, etc.) and, hence, was subjected to a different proportion of shares. Similar to an informal working group in paddy cultivation (*balambiak hari*), working in a team aimed not only at maximizing the chance for harvesting more fish by operating with particular fishing equipment. There was also a consciousness among team members (especially the group leader) that by working as a team they also shared the possible risk of incurring a loss. Risk aversiveness is a common practice in a community whose livelihood security is susceptible to extremely fluctuating natural resources availability, especially as the environments pose certain harmful threats to humans (Moran, 1982).

4.3.2 Livelihood Assets

This qualitative case study identified tangible and intangible resources fishermen utilized and managed to develop strategy to provide food and cash for their family, or livelihood assets. To describe those assets, this study employed the terminology and concepts in the sustainable livelihood approach (SLA), by which livelihood assets are classified into natural, human, social, financial, and physical capital.

4.3.2.1 Natural Capital

Natural capital refers to all extractable natural resources (goods and services) available from which fishermen may provide basic needs for their families; for example, land, coastal water, water, biodiversity, and so forth. In this study, availability did not refer to accessibility. There were local institutions[15] (rules and norms) through which natural capital was managed by a set of property right regimes, particularly to distribute land and coastal water to the locals. It was only through these institutions that land became accessible by the local people for utilization and management.

According to local customary rules, land in Nagari Sungai Pisang was regulated by two property right regimes: *ulayat kaum*[16] and *ulayat nagari*. *Ulayat nagari* referred to land owned by the *nagari*, which simply means owned by everyone who resides in the *nagari*. In Nagari Sungai Pisang, *ulayat nagari* might be distributed to anyone who resided in the *nagari*. To be granted with the right to access and utilize the land, one needed to propose to a customary council called *Badan Musyawarah*. *Ulayat nagari* comprised lowland forest area and cutch plantation (see areas number 1 and 2 in Fig. 4.2). Coastal water with all extractable resources (fish, coral reef, and mangrove forest) and the small islands were also part of *ulayat nagari*.

---

[15] Institutions have a broad range of definitions. Among the definitions that encompass most contemporary views of institution is the one provided by Scott (1995). Scott defines institutions as cognitive, normative, and regulative structures and activities that provide stability and meaning to social behavior. Vatn (2005)" has been changed to match the author name/date in the reference list. Please check here and in subsequent occurrences, and correct if necessary. Vatn (2005) differentiates institutions into conventions, norms, and externally sanctioned rules. However, Jentoft (2004) argues that in developing a definition of institutions and their constructions, one should include social and cultural aspects and not refer only to rational choice theory, especially in fisheries contexts. One of the latest studies that redefines institutions and their local constructions in capture fisheries by taking into account local contexts was carried out by Deswandi (2012) in Indonesia. In his thesis, he redefines and reconstructs institutions as conventions, norms, and rules by specifically taking into account types of punishment applicable in the prevailing social and cultural contexts.

[16] *Ulayat kaum* referred to land communally owned by tribes or *kaum*. Land was distributed to and managed by families that were the legitimate members of the tribe, to support the welfare of the tribes' members. *Ulayat kaum* was within and between families (and through generations) through a matrilineal inheritance line, which means only females of the tribe have the right to manage and utilize *ulayat kaum*. In Nagari Sungai Pisang, *ulayat kaum* mostly comprised paddy field, housing areas, and mixed-perennial plantation field locally called *parak*. (For deeper and more comprehensive reference for *ulayat* in West Sumatra, please see Benda-Beckmann and Von Benda-Beckmann, 2001.)

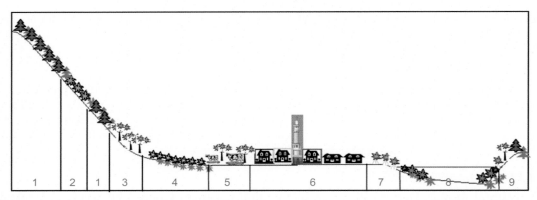

FIG. 4.2   A transect sketch of Nagari Sungai Pisang drawn from the east to the west side of the area. (The length, depth, height, and the number of objects (tree, cow, and house) do not represent the actual size and quantity.) 1, low-land forest vegetation; 2, cutch plantation; 3, perennial crops; 4, paddy field; 5, grazing land mixed with perennial crops; 6, areas of housing and public facilities; 7, mangrove and swamp areas; 8, sea grass areas and coral reefs; and 9, small islands with perennial crops (coconuts), shrub, and secondary mangroves.

4.3.2.2 Human Capital

Human capital refers to skills, knowledge, experience, and good health retained by individuals that might be empowered to create activities for food and cash. In Nagari Sungai Pisang, learning ability through observation (ie, social learning) was another human capital that enhanced individual adaptability imitating process (Moran, 1982; Bell et al., 1978). Accumulated memory of past experiences (remembrance) was a useful asset that supported the development of coping strategies in the present time (Berkes et al., 2003). Through social learning, which is imitating, fishermen learned paddy cultivation, boat making, and other alternative productive activities. Some fishermen became involved in paddy cultivation at a younger age, when helping their parents or relatives. Accumulated past experience has provided a means for reproduction of paddy cultivation.

Persistence, which lies within an individual's psychological constitution, plays an important role in helping individuals to cope with livelihood uncertainty. In Nagari Sungai Pisang, such psychological constitution prevailed and appeared to stem from a dominant religious belief/thought embrace by fishermen. In their faith, fishermen believed many uncontrolled/unexpected events were predetermined by God, including fishing uncertainty. Living with fishing uncertainty, fishermen must retain a quality of persistence to mentally cope with a low level of catch and still continue fishing on the other days and/or to create alternative activities.

4.3.2.3 Financial Capital

Financial capital was the most scarce livelihood asset owned by fishermen in the *nagari*. This should be inherent with the magnitude of economic activities in the *nagari* where almost all livelihood activities served mainly for subsistence. The forms of financial capital owned by fishermen in the *nagari*, hence, were also limited to cash saving, gold, and credit. Only a few fishermen obtained remittances.

4.3.2.4 *Physical Capital*

The most critical physical capital owned by the fishermen was fishing equipment and materials, such as boats, ships, fishing nets, trawl, and hand line. These were their major production factors. Theoretically, the more physical capital fishermen have the more likely the fishermen will develop and perform alternative productive activities. However, this was not always the case in Nagari Sungai Pisang. Not having physical assets did not simply mean a burden to developing livelihood strategies.

4.3.2.5 *Social Capital*

Social capital refers to the quantity and quality of the social network one has. Such a network is included in the membership in particular groups, as social relations (kin and quasi kinship). A social network matters. In Nagari Sungai Pisang, and in the context of livelihood strategies, a social network determines what you could ask for or borrow, who you could ask for help (and how many individuals you could ask), and when you could ask for help. Tribe, kin and quasi kin, membership in informal community groups (cash-saving and/or collaborative paddy cultivation working groups), and the practice of reciprocal generosity defined the extent of the social network fishermen could have. Having more individuals in their networks means more sources of help and support they could access in difficult times.

4.4 LIVELIHOOD STRATEGIES: BETWEEN THEORETICAL APPROACH AND EMPIRICAL CASE

The SLA (see, for example, CARE, 2002) is based on an assumption that the wider the range of livelihood assets households could have or access, the more likely it is those households will sustain their livelihoods. This qualitative case study can't provide a confirmation to such an assumption. However, by livelihood strategies, fishermen in Nagari Sungai Pisang have indicated their capacity and quality to cope with livelihood uncertainty that stemmed from artisanal fisheries and common-pool resource dilemmas in capture fisheries. Such capacity and quality were confirmed by a number of strategies and alternative activities they developed and performed. To that extent, this case study has rejected one common assumption that fishermen in Indonesia are season-dependent and rely on credit during the tropical monsoon season (see Dahuri, 2000). The localities in social and ecological circumstances in Nagari Sungai Pisang, this study would argue, have created a different outcome in the way fishermen responded to livelihood uncertainty.

In Nagari Sungai Pisang, the capacity to cope with livelihood uncertainty also involves a unique relationship between livelihood assets. Each livelihood asset did not simply serve as a complement to the existing assets. Interestingly, particular livelihood assets might substitute for other assets. Social capital has been a very useful livelihood asset for fishermen. Through a social network, fishermen could have access to utilize land (natural capital) and borrow money or take debts from relatives or local merchants (financial capital). Fishermen could also borrow fishing or farming equipment (physical capital) in certain situations. Through a social network, fishermen could easily reduce the need for financial capital in performing livelihood activities; for example, through *balambiak hari* in paddy cultivation or establishing a fishing group.

As an illustration, fishermen could access the right to utilize *ulayat kaum* to carry out paddy cultivation without necessarily owning the required paddy field. In this particular case, fishermen established cooperation with those who have a paddy field but for some reason did not utilize the land. Such cooperation was framed by a customary rule called the *patigoan* or "one-third" system. Under this "one-third," fishermen invested all inputs in paddy cultivation. Fishermen, at the end, would receive two-thirds from total harvested crop (gross) and the owner of paddy field would receive the rest (one-third) of the harvested crop. To establish this kind of cooperation, fishermen could directly propose to those who were granted with *ulayat kaum*. As cooperation was not officially written, trust and good reputation have served as the only warranty. Such cooperation might be renewed or ended only after the harvest. To maintain such mutual cooperation, both parties would need to comply with the existing customary rules and norms. While economic sanctions (ie, fines) could be enforced by the customary council to violators, social sanctions that automatically applied following the first sanction were perceived to be more costly for violators. Social sanctions could be enforced by almost all individuals in the *nagari*. This could mean a timeless social exclusion and, hence, the end of possible support from others. In a socioeconomic context, where reciprocal generosity and dependence on a social network could be the most critical measure for livelihood survival, permanent exclusion from a group and/or the whole community would be unbearable (Moran, 1982). On the other hand, this would also serve as a disincentive for compliance with rules and norms (Jaffe and Zaballa, 2010; Gürerk, et al., 2006; Henrich, 2006; Posner and Rasmusen, 1999). With social capital prevailing as the most prominent asset that could substitute for the absence and/or compensate for the lack of other livelihood assets (natural, human, physical, and financial capital), livelihood strategies performed by fishermen in the *nagari* was a collective social phenomenon.

4.5 CONCLUSION

The term *rasaki harimau* describes how the fishermen in Nagari Sungai Pisang perceived fishing uncertainty as a part of their daily lives. To secure their families' livelihoods, they developed livelihood strategies as coping mechanisms for fluctuating and unpredictable fish catches that might create livelihood insecurity by various activities. The strategies ranged from relying on the kindness of nature, alternative productive activities, forecasting for unexpected future events, maintaining or enhancing social relationships, and risk-spreading mechanisms and productivity enhancement. Diversification of livelihood strategies and activities served as redundancy for the fishermen. Through redundancy, fishermen created backup activities. In a case where one activity failed or partially failed to provide food and cash for their families, fishermen might expect to fulfill subsistence need and hopefully some extra saving from other activities. Paddy cultivation was considered as the most important activity under the strategy developing alternative productive activities, as this activity could provide families with rice. To develop livelihood strategies, fishermen relied on ownership and/or access to utilize livelihood assets that were classified into natural, human, financial, physical, and social capital.

Natural, human, financial, and physical capital happened to be limited livelihood assets (in terms of access and availability) in the *nagari*. However, fishermen who possessed extensive social capital (as manifested in an extensive social network) might have the chance to access those scarce assets. Such access to utilize livelihood assets owned by tribes (the *ulayat kaum*) and nagari (the *ulayat nagari*) was managed by local customary rules. The norm of reciprocal generosity has served as an incentive to promote cooperation within the community. The customary institutions (rules and norms) have promoted and facilitated redistribution and optimum utilization of critical and limited livelihood assets within the *nagari*. On the other hand, rules and norms also served as social control (through economic and social sanctioning) to ensure compliance with the commitment for mutual cooperation.

The social network appeared as the most resourceful and functioning livelihood asset among fishermen. Livelihood strategies were successfully developed and performed by fishermen; therefore, it could be concluded as the results of collective actions performed by the community in the *nagari*. This perspective might contribute to the development and adoption of an alternative approach (ie, methodology) to understanding sustainable livelihood and livelihood security, through livelihood strategies. The three concepts might be perceived as social phenomena that emerges from collective interaction among intelligence individuals (who store cumulative past experience as the basis for future decision making) whose actions/decisions are driven/motivated by a few simple rules of interaction by which individuals independently act and/or make decision to maximize self-benefits (which range from pure altruistic to pure economic) and minimize or avoid loss and punishments (economic and social sanctions) (see Johnson, 2009; Folke et al., 2007; Lansing, 2006; Waldrop, 1992). By weighting the potential cost and benefit, an individual might decide whether to cooperate (ie, sharing livelihood assets through a social network) or not to cooperate. Employing this approach, qualitative and quantitative research on livelihood strategies and sustainable livelihood might employ computation/modeling; for example, through agent-based modeling that focuses on emergence phenomena from interactions between resource users, natural resources, and the market under given institutions (Johnson, 2009). By doing so, knowledge constructed from a qualitative case study could be useful to other cases in different social, economic, cultural, and biophysical environments (Donmoyer, 2008; Berkes, 2006).

References

Adger, W.N., 2000. Social and ecological resilience: are they related? Prog. Hum. Geogr. 24 (3), 347–364.

Afrizal, 2005. Pengantar Metode Penelitian Kualitatif: Dari Pengertian Sampai Penulisan Laporan. Laboratorium Sosiologi FISIP, Universitas Andalas, Padang.

Bell, P.A., Fisher, J.D., Loomis, R.J., 1978. Environmental Psychology. W.B. Saunders Company, Toronto.

Berkes, F., 2006. From community-based resource management to complex system: the scale issue and marine commons. Ecol. Soc. 11 (1), 45 (accessible under). http://www.ecologyandsociety.org/vol11/iss1/art45/.

Berkes, F., Colding, J., Folke, C. (Eds.), 2003. Navigating Social-Ecological System: Building Resilience for Complexity and Change. Cambridge University Press, Cambridge.

Biro Pusat Statistik, 2004. Kecamatan Bungus Teluk Kabung dalam Angka. Biro Pusat Statistik, Padang.

CARE, 2002. Household Livelihood Security Assessments: A Toolkit for Practitioners. Prepared for the PHLS Unit by TANGO International Inc., Tucson, AZ.

Creswell, J.W., 2009. Research Design: Qualitative, Quantitative, and Mixed Methods Approaches. Sage Publications, Inc., Los Angeles, CA.

Dahuri, R., 2000. Pendayagunaan Sumber Daya Kelautan Untuk Kesejahteraan Rakyat. LISPI, Jakarta.

Davidson-Hunt, I.J., Berkes, F., 2003. Nature and Society through the Lens of Resilience: Toward a Human-in-ecosystem Perspective. In: Berkes, F., Colding, J., Folke, C. (Eds.), Navigating Social-Ecological System: Building Resilience for Complexity and Change. Cambridge University Press, Cambridge, pp. 53–82.

Deswandi, R., 2012. Understanding institutional dynamics: the emergence, persistence, and change of institutions in fisheries in Spermonde Archipelago, South Sulawesi, Indonesia. Doctoral thesis. Faculty of Social Science, University of Bremen, Bremen.

Donmoyer, R., 2008. Generalizability. In: Given, L.M. (Ed.), The Sage Encyclopedia of Qualitative Research Methods. Sage Publications, Inc., Thousand Oaks, CA, pp. 371–372.

Folke, C., Colding, J., Berkes, F., 2003. Synthesis: building resilience and adaptive capacity in social-ecological systems. In: Berkes, F., Colding, J., Folke, C. (Eds.), Navigating Social-Ecological System: Building Resilience for Complexity and Change. Cambridge University Press, Cambridge, pp. 352–387.

Folke, C., Lowell Jr., P., Berkes, F., Colding, J., Svedin, U., 2007. The problem of fit between ecosystems and institutions: ten years later. Ecol. Soc. 12 (1), 30 (accessible under). http://www.ecologyandsociety.org/vol12/iss1/art30/.

Glaser, M., Radjawali, I., Ferse, S., Glaeser, B., 2010. 'Nested' participation in hierarchical societies? Lessons for social-ecological research and management. Int. J. Soc. Sys. Sci. 2 (4), 390–414.

Gürerk, Ö., Irlenbusch, B., Rockenbach, B., 2006. The competitive advantage of sanctioning institutions. Science 312 (5770), 108–111.

Henrich, J., 2006. Cooperation, punishment, and the evolution of human institutions. Science 312 (5770), 60–61.

Jaffe, K., Zaballa, L., 2010. Co-operative punishment cements social cohesion. J. Artif. Soc. Soc. Simul. 13 (3), 4.

Jentoft, S., 2004. Institutions in fisheries: what they are, what they do, and how they change. Mar. Policy 28, 137–149.

Johnson, N., 2009. Simply Complexity: A Clear Guide to Complexity Theory. Oneworld Publications, Oxford.

Lansing, S.J., 2006. Perfect Order: Recognizing Complexity in Bali. Princeton University Press, Princeton, NJ.

Lengkong, F., 2007. The psychological impact of disaster: the risk factors associated with posttraumatic stress of the 2004 tsunami survivors in Banda Aceh city and Aceh Besar district. In: Paper Presented in the International Symposium Disaster in Indonesia: Problems and Solutions on July 26–28th 2007 in Padang, Indonesia.

Low, B., Ostorm, E., Simon, C., Wilson, J., 2003. Redundancy and diversity: do they influence optimal management? In: Berkes, F., Colding, J., Folke, C. (Eds.), Navigating Social-Ecological System: Building Resilience for Complexity and Change. Cambridge University Press, pp. 83–114.

Moran, E.F., 1982. Human Adaptability: An Introduction to Ecological Anthropology. Westview Press, Boulder, CO.

Pelras, C., 2000. Patron-client ties among the Bugis and Makassarese of South Sulawesi. In: Tol, R., Von Dijk, K., Acciaioli, G. (Eds.), Authority and Enterprise among the Peoples of South Sulawesi, vol. 156 no. 3, KITLV Press, Leiden, pp. 393–432.

Pelras, C., 2006. Manusia Bugis (The Bugis). Nalar, Jakarta. A.R. Abu, Hasriadi and N. Sirimorok (trans).

Pet-Soede, C., Densen, W.L.T.V., Hiddink, J.G., Kuyl, S., Machiels, M.A.M., 2001. Can fishermen allocate their fishing effort in space and time on the basis of their catch rates? An example from Spermonde archipelago, Southwest Sulawesi, Indonesia. Fish. Manag. Ecol. 8 (1), 15–36.

Posner, R.A., Rasmusen, E.B., 1999. Creating and enforcing norms, with special reference to sanctions. Int. Rev. Law Econ. 19 (3), 369–382.

Sallatang, M.A., 1982. Pinggawa-Sawi: Suatu Studi Kelompok Sosiologi Kecil. Doctoral thesis. Universitas Hasanuddin, Makassar.

Schlager, E., Ostorm, E., 1999. Property rights regimes and coastal fisheries: an empirical analysis. In: McGinnis, M.D. (Ed.), Polycentric Governance and Development. University of Michigan Press, Ann Arbor, MI.

Scott, W.R., 1995. Institutions and Organizations. Sage Publications Inc., Thousand Oaks, CA.

Stainback, S., Stainback, W., 1988. Understanding and Conducting Qualitative Research. Kendall/Hunt Publishing Company, Dubuque, IA.

Vatn, A., 2005. Institutions and the Environment. Edward Elgar Publishing, Cheltenham.

Von Benda-Beckmann, F., Von Benda-Beckmann, K., Marks, H. (Eds.), 1994. Coping with Insecurity: An 'Underall' Perspective on Social Security in the Third World. Stichting Focaal, Centrale Reprografie, Universiteit van Nijmegen, Nijmegen.

Von Benda-Beckmann, F., Von Benda-Beckmann, K., 2001. Recreating the nagari: decentralization in West Sumatera. Max Planck Institute for Social Anthropology Working Papers No. 31, Max Planck Institute for Social Anthropology, Halle.

Waldrop, M.M., 1992. Complexity: The Emerging Science at the Edge of Order and Chaos. Simon & Schuster Paperbacks, New York, NY.

Yayasan Hayati Lestari, 2006. Rencana Strategis Pengelolaan Kawasan Pesisir dan Laut Sungai Pisang. Yayasan Hayati Lestari and Dinas Kelautan dan Perikanan Kota Padang, Padang.

Utilization Rights of Sikuai Island and Pasumpahan Island, West Sumatra: Study on Implementation of Community-Based Property Rights of the Local Community of Sungai Pisang Village

S.M. Sari*, N. Effendi[†], A. Saptomo[‡]

*Freelance consultant, Bukittinggi, Indonesia [†]Andalas University, Padang, Indonesia [‡]University of Pancasila, Jakarta, Indonesia

5.1 INTRODUCTION

Many small islands in Sumatra remain unoccupied due to their remoteness and limitations. These limitations start from the limitation of natural resources, limitation of capital to invest, and also limitation of human resources. The geographical condition of a small island makes the nature and ecosystem unique and often different with the mainland. These issues not only occur on unoccupied small islands, but also on the occupied ones.

In effect, the occupancy of an unoccupied small island can result to the loss of ownership rights of the small island.[1] The loss of the ownership rights of an island can be seen from three perspectives: economic, political, and legal (Suhana, 2005). The island can be said to be economically lost when it is being utilized or managed by another nation although, legally, that island belongs to Indonesia. Utilization or management rights can be gotten legally from the Indonesian government, for instance through a leasing agreement, or can be obtained illegally. From a political

---

[1] In effect, the occupancy of Sipadan and Ligitan islands was one of the reasons for the loss of ownership rights of the islands from the Republic of Indonesia to Malaysia.

point of view, an island can be said to be lost if the community itself preferred to admit the other nation as their government rather than their own government[2] (Suhana, 2005). Small islands can be viewed as lost also if there is a ruling from international law that states that the island no longer belongs to the nation that formerly owned it. This process takes longer to happen because it must go through extensive international discussion and even confrontation. Besides, it also needs scientific evidence that can show the existence of the island in the new country's territory.[3]

Generally, islands have long been noted for their unique fauna and flora, which are particularly vulnerable to disturbance and destruction by human activities. Islands are also of interest for the special adaptations of island societies, the difficulties of economic development in an island context, and the challenge of achieving sustainable development within limited island resources (Khaka, 1998; Mook, 1998; Annan, 2004; Shepherd, 2006). With the increasing global cimate change, islands represent some of the most fragile and vulnerable resources on the planet (UNEP, 2004). An island is any piece of land smaller than a continent and larger than a rock that is completely surrounded by water. Groups of related islands are called archipelagos.

Indonesia is the second biggest archipelago state in the world and consists of more or less 17,500 big and small islands (Dahuri, 2000). Many of its small islands are not well managed by the government, while in fact many natural resources also lie on those islands. Lately, the cases of small islands utilization have become a major concern. As a matter of fact, most Indonesian small islands have been utilized by their investors[4] and, therefore, it affected the utilization activities of the local community to the islands.

Usually, the islands that have been utilized/managed by the investors become prohibited areas for the local community. Because the investors made the islands their private domain, and in advance, they made a regulation that prohibits the local community from using the islands. This prohibition resulted in the surrounding community not being able to even enter an island that they used to go to in the past. In other words, local community access and control are also being affected. Conflict over the resource users (local community vs. the investors) might appear in this situation, and sometimes it cannot be avoided. By the rights that the investors have,[5] there is lack of choices that can be taken by the community as their community-based property rights (CBPRs) over the resources are not recognized by the government.

The resources on community-based property land are the rights of the local people who own the rights. The local people have the rights to extract the resources on the land and make

---

[2] We can take the case of Miangas Island as an example. Legally, the island belongs to Indonesia, but politically, it belongs to Philippine. Because the language the community used is the Tagalog language and the currency they used is the peso, Philippine's currency, not the rupiah.

[3] This was what happened to Indonesia in the case of the loss of Sipadan and Ligitan islands. Indonesia had struggled to defend these islands in many efforts (through diplomacy, etc.) for many years. The efforts that Indonesia made produced no results, so both Indonesia and Malaysia finally agreed to bring this case to the International Court of Justice (located in Den Haag, Netherlands) and the court ruled that Sipadan and Ligitan islands are under the Malaysian government's jurisdiction. Indonesia lost the case.

[4] The investors are often people that come from outside the community. Besides that, the investors can also be foreigners or government. In this research, these investors wer also called the outsiders.

[5] Most of the investors have evidence, which is legal, and can be used if the conflict is brought to court. This evidence is usually in the form of a piece of paper that states the investor now has the power to control and manage the island based on the agreement that the investor made with the person that has the power to lease the island. The person could be from the government or the community itself. That agreement paper also states the rights and obligations of both parties.

I. LIVELIHOOD DEPENDENCE, RIGHTS AND ACCESS TO NATURAL RESOURCES

optimal use of the land. The rights are transferred hereditarily from generation to generation without reducing the rights from the former to the future generation.

The transfer of the utilization rights from the local community to the investors has an impact on the local community. They are might not be able to withdraw resources from the island. They also might not be able to take advantage of the island. Or maybe they might not be able to enter the island to get the service that is provided by the island. There are many possibilities for impacts from the rights' transfer process that are still unknown, either positive or negative impacts.

The utilization rights that stick to the land of the community-based property have a significant role in the resources extraction activity of the local community. By leasing the land to outsiders, it means the rights of the community that stick to land are also leased to the outsiders. The utilization rights of a land by the community had a close relationship to the CBPRs of the community to the land. The community can take and extract the resources on the land only if the land is still the community-based property of the community where the utilization rights stick to the land.

This chapter explores the existence of small islands as CBPRs. The research took place at Sungai Pisang Village, geographically on E 1000 20′ 15″ to 1000 24′ 15″ and S 10 5′ 0″ to 10 8′ 15″; and administratively known as Kelurahan[6] Teluk Kabung Selatan. It is located in Bungus Teluk Kabung Subdistrict, Padang Municipality, West Sumatra. This area is 10 km from the su-district and 31 km from the municipality (BPS, 2006). The total population of Sungai Pisang Village is 1714 individuals, conprising 851 males and 863 females (Anonymous, 2007).

There are six small islands included in the territorial area of this village. Two of them are now utilized by investors: Sikuai Island and Pasumpahan Island. Sikuai Island became a tourism resort and Pasumpahan Island became a vessel workshop.

The existence of investors in both islands results in a reduction of rights that the local community holds. The investors create regulations that restrict the local community's rights to the island. As the investors hold the valid rights to the island (according to the positive law), the local community has to obey the regulations issued by the investor. The leasing process of the islands was made based on *adat* law and positive law. Therefore, the involvement of *adat* leaders and government officers is found in this case. Later, this chapter will describe the utilization rights the local community had before and after the utilization activities by the investors, the implementation of the local community's rights, problems during rights implementation, and a problem-solving mechanism.

5.2 LITERATURE REVIEW

5.2.1 Defining Community-Based Property Rights

Property rights are an important factor in natural resources management. Property rights determined the rights to manage and control natural resources. By having a property right over a specified area, one can explore the resources beneath and upon it, manage the use of the resources, and extract the resources at will.

---

[6] Kelurahan is an area that is administratively lower than the subdistrict level. The official head of a Kelurahan is an officer called Lurah. The Lurah officer (government officer who chairs at the Kelurahan level) was one of the local people of Nagari Sungai Pisang. In the Kelurahan office, he was helped by a secretary and three staff members. All of them are also local people of Sungai Pisang.

Indeed, sustainable development is unlikely to ever be attained in many locales if the property rights of indigenous and other local communities remain unrecognized by national and international laws (Lynch and Chaudhry, 2002). CBPRs by definition derive from and are enforced by communities. The distinguishing feature of CBPRs is that they derive their authority from the community in which they operate, not from the nation-state where they are located. Formal legal recognition or grant of CBPRs by the state, however, is generally desirable and can help to ensure that CBPRs are respected and used in pursuit of the public interest.

5.2.2 Legal Recognition of Community-Based Property Rights

Traditional villages, local communities, and CBPRs are typically heterogeneous and dynamic, and many exist and function outside of official government structures. One useful conceptual tool for clarifying these facts is to distinguish between the *grant of legal rights* by the state and the *legal recognition of CBPRs*. Legal rights do not emanate only from nation-states. There are various theories of law and jurisprudence that acknowledge as much. When national governments own land and other natural resources, they can decentralize authority to local government units or local officials, which then grant management/property rights to communities located within their jurisdiction. But when CBPRs already cover an area, the state may (and often should) be obliged to recognize these rights, especially when the area is an ancestral domain/indigenous territory that predates the state and its natural resource classifications.

Decentralization can help foster and support legal recognition of CBPRs and various types of community-based natural resources management (CBNRM) initiatives[7] but decentralization to local government units does not necessarily lead to such outcomes.

There are many other reasons for legally recognizing CBPRs. First and foremost, in many countries the constitution can be interpreted as already protecting the CBPRs of indigenous peoples (ie, original long-term occupants). Legally recognizing these rights would be a positive and crucial step toward ensuring that the constitution is invoked to protect and promote the well being of all citizens. In many countries where conflict is epidemic, the legal recognition of CBPRs would also contribute to goodwill between local communities and governments (Lynch and Harwell, 2002).

[7] For details, see Rahmadi (2004). The Indonesian Constitution of 1945 legally recognizes the *adat* community. The notion of *adat* community specifically refers to those who have the rights of origins *(hak asal usul)*. The rights of origins are recognized by the Constitution of 1945 prior to and after the amendment established in Article 18 that states: "In the Indonesian territory there are about 250 autonomous communities… The Indonesian government respects the status of them and the state law pertain them will take into account of their rights of origins." The legal recognition of the *adat* community in the Constitution of 1945 after the Second Amendment is established in Article 18 B. The Forestry Act of 1967 stipulates that the rights of *adat* community to utilize forest are recognized "in so far in reality they still exist" ("*Sepanjang menurut kenyataannya masih ada*"). In order to utilize the rights to forest, therefore, the *adat* community should be declared, first, by the District Head *(Bupati)*. This provision is stipulated in Decision No. 251 of the Minister of Forestry of 1993 Article 2, which grants the district head *(Bupati)* the power to declare whether or not *adat* communities in his or her territory still exist. The more power that the central government transferred to local government, the greater the possibility that the local community's CBPRs will be recognized.

Many rural people in Indonesia and throughout the developing countries are guardians and stewards of natural resources, including biodiversity reservoirs and carbon sinks, and possess important local knowledge for managing these resources sustainably. Most property rights theorists and students rely on a four-part typology: private (which is a misnomer because it really means individual), commons, state, and open access (which refers to a situation where no defined property rights exist) (Lynch, 1999). This typology has proven to be very useful in distinguishing common property from open access. Therefore, Lynch comes out with the idea of property rights using two conceptual and interrelated spectra, as follows:

1. The first spectrum has the public on one end and private on the other. Public means property owned by the state and private means property not owned by the state. However, the degrees of private and public ownership vary with some private rights being heavily burdened by state conditionalities such as easements and zoning restrictions, and some public rights being largely unregulated. Private titles, therefore, are not necessarily the strongest type of property right. To know what a specific property right requires, whether it's a private title or a public lease, requires that you identify what's in its bundle.
2. The other spectrum has the individual on one end and the group on the other. The group end basically refers to CBPR regimes, most of which typically include individual rights as well as common properties. Individual property rights do not only emanate from the nation-state. Rice terraces in Southeast Asia provide a good example of this. A rice terrace in the middle of a public forest zone in Asia that wasn't individually owned, is usually in accordance with a local CBPR regime. Cross-referencing the two spectra allows for the identification of four types of possible property rights, as discussed below.

5.2.2.1 *Private-Individual Property Rights*

A private-individual property right is widely believed to be the best possible type of property right, because individual private owners tend to have the greatest freedom to use resources as they see fit.

5.2.2.2 *Public-Individual Property Rights*

This type of property right typically refers to a lease to an individual (or a corporation, which in some nations is deemed to be the legal equivalent of an individual) of legal rights over land that is ostensibly government owned. Timber concessions are a classic example. Over the past decade in Asia, as the plight of rural people has garnered more attention, several countries such as Thailand and the Philippines have also developed programs for conditionally leasing public-individual property rights to small-scale forest farmers, usually for a fairly short period of time ranging from 2 to 25 years.

5.2.2.3 *Public-Group Property Rights*

This category refers to legal arrangements where a government conditionally leases or otherwise delegates property rights, usually for a specific period of time, to a local community or user group. With the growing emphasis on such concepts as community forestry and CBNRM the range of new types of public-group property rights is increasing and would include the Communal Areas Management Program for Indigenous Resources

(CAMPFIRE) program in Zimbabwe (which is still limited to wildlife) as well as the Joint Forest Management program in India.

5.2.2.4 Private-Group Property Rights

The best (although the rarest and most difficult to acquire) option for protecting CBPPs, including the commons, especially for original long-term occupants of a specific area, would be to acquire legal recognition of private-group rights, a concept that should encompass individual and common property rights within the perimeter of a community-based property regime (Fig. 5.1).

5.2.3 Elements of Property Rights

Property rights are the entitlements defining owners' rights and duties in the use of the resource. This consists of several elements that built the property rights, which are (Shivakoti, 2006):

a. Access: the right to enter a defined physical property.
b. Withdrawal: the right to withdraw the product of the property; for example. harvest fuel wood, catch fish, use water, and so forth.
c. Management: the right to regulate internal use and make improvements in the resource.
d. Exclusion: the right to determinate who will have an access right and how that right may be transferred to others.
e. Alienation: the right to sell or lease the collective choice rights of management and/or exclusion.

From the five elements above, access is the lowest level and alienation is the highest level of right hierarchy of a property. The right that can be gotten from access is the least right

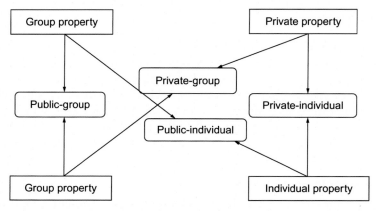

FIG. 5.1   Spectrum of property rights. *Lynch, O.J., Harwell, E.E., 2002. Whose Resources? Whose Common Good? Towards a New Paradigm on Environmental Justice and the National Interest in Indonesia. Center for International Environment Law (CIEL) in collaboration with Perkumpulan untuk Pembaruan Hukum Berbasis Masyarakat dan Ekologis (HuMa), Lembaga Studi dan Advokasi Masyarakat (ELSAM), Indonesia Center for Environmental Law (ICEL), International Centre for Research in Agroforestry (ICRAF), Jakarta.*

that one can have from a property. At the access point, one may enter a defined physical property, but may not take something from it. If one has the withdrawal right, he may enter the property and take the benefits from it. A management right holder of a property may take the benefit of the property and may participate in managing and be responsible for that. As the bundle of rights increases in each level, the holder of the right has the power or the authority upon the property in every decision making of the property. Access and withdrawal are operational-level property rights, while management, exclusion, and alienation are collective-choice-level property rights.

Based on those rights of property, the holder of each right can be categorized into positions in property rights (Schlager and Ostrom,1999 in Shivakoti, 2006). It is divided into the four groups discussed next.

5.2.3.1 Authorized Users

The individuals who are holding the rights of access and withdrawal are called authorized users. These rights may be transferred to others either temporarily or permanently. Their rights are defined by others who hold collective-choice rights of management and exclusion. In this case, they cannot devise their own harvesting rules.

5.2.3.2 Claimants

Claimants are the individuals who possess operational rights and the collective-choice rights of management. They cannot specify who may not have access to resources nor can they alienate their right of management.

5.2.3.3 Proprietors

Proprietors are the individuals who possess collective-choice rights to participate in management and exclusion.

5.2.3.4 Owners

Owners are the individuals who can sell or lease their collective-choice rights (ie, rights of management and/or exclusion).

These positions also have the hierarchy, with the authorized users at the lowest point and owner point. The combination between bundles of rights associated with the position in the property rights regime can be concluded in Table 5.1.

TABLE 5.1 Bundles of Rights Associated With Position

Right/Position	Owner	Proprietor	Claimant	Authorized User
Access & withdrawal	X	X	X	X
Management	X	X	X	
Exclusion	X	X		
Alienation	X			

Source: Schlager and Ostrom (1999) in Shivakoti, G., 2006. Lecture Note on Natural Resources Management and Planning and Policy, at Integrated Natural Resource Management Study Program. Post Graduate Andalas University, West Sumatra, Indonesia.

5.2.4 Stage of Conflict

Before reaching the conflict stage, the conflict itself walks through some processes. The process of conflict includes the stages of conflict potential (conditional), preconflict (monadic), conflict (dyadic), and dispute (triadic) (Saptomo, 2006). Every step in the process of conflict has its own alternative dispute resolution that includes adjudication, arbitration, mediation, negotiation, or without the help of a third party, which is by coercion, conquest, avoidance, and lumping it (Nader, Laura and Harry F. Todd, Jr., 1978, in Saptomo, 2006).

According to Nader and Todd (1978) as cited from Saptomo (2006), before a dispute happens, a package of conditions level is advances, as discussed below.

5.2.4.1 Preconflict

Preconflict is the condition that shows the social relationship between two parties or more, but one of the parties feels being disadvantaged by the other party so that condition makes the disadvantaged party grievances. This step is known as monadic conflict.

5.2.4.2 Conflict

Conflict is a condition that shows the social relationship between two parties or more, where the disadvantaged party complains about his grievances to the party that caused the disadvantage. This condition involves two parties in a conflict, and it is also known as dyadic conflict.

5.2.4.3 Dispute

This is the condition where it has reached the dispute level between the parties, and a third party is often involved in the conflict. Later, this condition can be called a triadic conflict.

Conflicts are multidimensional and frequently involve complex interactions between many parties. However, for analytical purposes it is useful to identify the following four dimensions of a conflict: the actors; the resource in dispute; the stake that each actor has in the resource; and the stage that the conflict has reached (ie, the time dimension). An environmental dimension will be added to each of these (Scialabba, 1998).

5.2.5 Conflict Resolution

There are a number of ways of dealing with a conflict, ranging from violence at one extreme to ignoring the conflict at the other, with a variety of approaches in between. Toward the more hostile end of the spectrum is litigation, in which parties take their grievances to a court or tribunal, which applies predetermined legal rules to the conflict and issues a decision that is binding upon the parties, producing a winner and a loser (Scialabba, 1998).

However, parties are turning increasingly to alternative dispute resolution techniques to settle their disputes. These include negotiation, mediation, and conciliation, which are more flexible and produce results that are more acceptable to the parties as well as more sustainable in the longer term. Alternative dispute resolution is being used increasingly in conflicts over the environment and natural resources and has considerable advantages over traditional contentious methods (Scialabba, 1998).

Some analysts have used the concept of a mountain to symbolize the range of options faced in managing conflicts and explained it as follows:

"At the summit of the mountain is cooperative teamwork, with the goal of achieving a synergy of solutions of mutual advantage to all interests. At the base of the mountain, from where any climb has to begin, are isolation, the decision not to engage in the debate at all, and confrontation, in which positions have been adopted in fixed opposition to one another" (Brown et al., 1995 in Scialabba, 1998). From isolation and confrontation at the base of the pyramid the options progress through the stages of litigation, arbitration, mediation, facilitation, conciliation, negotiation, and on to cooperation at the top (Fig. 5.2).

5.2.6 Customary Rights in *Minangkabau* Concept

Management of natural resources in *Minangkabau* is regulated in the customary law concept. According to the lesson of customary law in *Minangkabau*, a customary right is the power or the authority that is owned by the customary law community upon a certain area or space; where all the members of the community are allowed to enjoy the advantage of the natural resources to sustain their livelihood, and it derives from the physical and spiritual hereditary relationship from ancestors to the present generation and also future generations (Narullah, 1999 in LBH Padang, 2005). It can be concluded that the customary right is the highest authority right upon an area and space in *Minangkabau*, whether it is about the space at the lowest level in a family or the space that is owned communally by the community, like the *nagari*[8] (LBH Padang, 2005).

A customary right in *Minangkabau* covers several aspects, including water, land, and air. In daily practice, the customary right is often related to customary land; sometimes people use the term *customary land* for defining a customary right, just because they assume that water (except the ocean) and air cannot be separated from land. In conclusion, it can be said that customary land is the most prominent object in the *Minangkabau* natural resources management context; besides, the customary rights upon water (river and ocean) and air are still untouchable (LBH Padang, 2005).

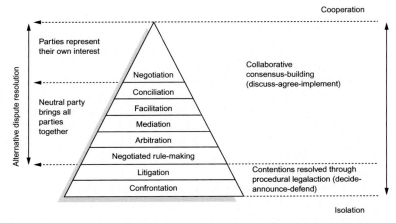

FIG. 5.2   Managing conflict. *Scialabba, N. (Ed.), 1998. Integrated Coastal Area Management and Agriculture, Forestry and Fisheries. FAO Guidelines. Environment and Natural Resources Service. FAO, Rome.*

[8] *Nagari* means village.

Basically, a customary right in the *Minangkabau* community, aside as the asset or wealth in economic perspective, has several noble functions, which are (Naim, 1984 in LBH Padang, 2005) as follows:

1. The right that has been accepted hereditarily from the ancestors who built the *nagari*; historical and religious characterized.
2. Equal right to all members of the community or the whole community; social justice characterized.
3. Customary right is not just for present generations but also the rights of future generations; sustainable characterized.

5.3 METHODOLOGY

5.3.1 Approach

This research explores how the Sungai Pisang Village local community is optimizing the utilization rights of the small islands as their CBPRs. Two of the six small islands that the community has are already leased to the investors: Sikuwai Island and Pasumpahan Island. The changing condition of the two islands from unleased islands to leased islands has caused the imitation of rights for the community to utilize the islands. Therefore, this research will be carried out in the qualitative method, as this is a proper method to have a better understanding of how the community can adjust due to the changing condition (Stainback and Stainback, 1988). The assumptions that support the choice of the qualitative method as the research method of this research are the research problems of this research demands a proper understanding of the real situation on the field as there is a lack of theory and previous research on this topic (Morse, 1999 in Creswell, 2003); this research puts greater emphasis on process rather than the product or result (Merriam, 1988 in Creswell, 2003).

5.3.2 Selecting Key Informants and Informants

There were eight key informants interviewed during the field work for this research. They were *adat* leaders, government officers (Lurah), and institutional leaders. They were eligible because they were directly involved in every decision-making process within the community; especially in the leasing agreement of Pasumpahan Island where they were the main actors that represented the community. The *adat* leaders were the Head of Badan Musyawarah and the Penghulu of four clans (Caniago, Koto Piliang, Jambak, and Tanjung). The institutional leaders were the leaders of the institutions that exist in the community of Nagari Sungai Pisang; they were the Head of Community Empowerment Institution and the Chairman of Youth organization.

Besides that, there were eight informants in this research. The composition of the informants was four males and four females. The occupation of the male informants is fisherman. Two of the women informants are housewives, while the other two are widows and farm for a living. The Lurah of Sungai Pisang was also invited as the representative of the Indonesia government, particularly the Municipality of Padang government in the research location jurisdiction.

5.3.3 Research Instrument

In this research, the researcher is the main research instrument, both for data collection and data analysis (Merriam, 1988 in Creswell, 2003). This is because a human being is in the best position to grasp the meanings people give to the events in their life. Human possess sufficient adaptability to encompass and adjust to the variety of realities that will be encountered when doing qualitative research in natural settings (Guba and Lincoln, 1981 in Stainback and Stainback, 1988). Other research instruments used in this research are research question guidance, video recorder, voice recorder, camera, and notes. Research question guidance is helping to keep the interview session in track and it can be used as a security blanket when the mind goes blank in the middle of an interview (Gaskell, 2000).

5.3.4 Data Collection Techniques

During May 2007 to April 2008, the field work for collecting data process was conducted, both for primary and secondary data. Nagari Sungai Pisang was periodically visited every two or three months, depending on the availability of the key informants, because not all of the key informants were available for an interview in the same day of visiting. On every visit, the researcher stayed three to four days within the community.

All data relevant to this research were classified as primary and secondary data. The sources of primary data were obtained from key informants and informants. The collecting data techniques were semistructured in-depth interviews and participant observations. Besides semistructured in-depth interviews, another type of interview was developed, dubbed as "light conversation." This type of interview was used when the researcher was facing the informants. Therefore, it is considered one of the techniques used for data collecting. There was much valuable information gained from these conversations that can be used to add the information gained from key informants. This technique was also used to cross-check whether what the key informants said matched the perspectives that live in the community.

The semistructured in-depth interview was carried out based on a set of main relevant questions compiled in an interview guideline.

The interview activities were conducted at the key informants' house during their spare time or after making an appointment, considering the business of the key informants. The participant observation was conducted during the *Tulak Bala*[9] ritual. The researcher might join

---

[9] This is a ritual held by the local community to eliminate the bad fortune from the *nagari*. This ritual covered three stages: (a) *Manawa*, an activity where all the community members come to the mosque (where Angku Bujang Sabaleh will stay during the ritual) and bring a bottle of water so that Angku Bujang Sabaleh can put a spell on the water to cure the disease that they have. (b) *Mandarah*, an activity done the day after the arrival of Angku Bujang Sabaleh, which is conducted at the upper course of the water resource that waters the entire area of the *nagari*, at a place called *Kapalo Banda* by the local community, where a goat is butchered and the blood of the goat drops to the water and flows, a ritual the community believes will eliminate bad fortune by spilling the blood of the sacrificed animal. (c) *Maratik*, the last activity conducted in this ritual. is divided into two parts: *maratik* on the ground and *maratik* on the ocean. The blood of the butchered goat, besides being dropped into the river, is also patched in a bucket. This blood is then mixed with the water from *Kapalo Banda*. Later, this blood is sparkled to the entire area of the *nagari*. During the sparkling activity, the escort is humbling a *salawat*.

the process of the *Tulak Bala* ritual, from the beginning to the end.[10] The ritual started with the arrival of Angku Bujang Sabaleh[11] and ended by *maratik* to the ocean that covered the jurisdiction area of Sungai Pisang. The route of *maratik* at the ocean reaches to Pasumpahan Island and Sikuai Island.

Data gained from key informants were the main data in this research; while the data gained through light conversations[12] with the informants were used as additional and supporting data to the main data in this research. From participant observation,[13] the data on the existence of the islands (especially Sikuwai Island and Pasumpahan Island) to the local community of Nagari Sungai Pisang is gained.

The secondary data were in the form of government documents, maps, regulations, and reports on previous relevant studies to the research topic. Those data were collected from several related governmental offices and nongovernmental organization: the Law Division of the Mayor's Office, Kelurahan Office, Central Bureau of Statistics, Ocean and Fishery Department of Padang Municipality, National Agrarian Agency of Padang Municipality, Tourism and Culture Department of Padang Municipality, Yayasan Hayati Lestari, and PSI-SDALP Library at *Andalas* University. Besides that, there were also electronic data derived from Internet sources.

5.4 UTILIZATION RIGHTS IN SIKUAI ISLAND AND PASUMPAHAN ISLAND

Basically, the common *nagari* customary rights (or in general known as CBPRs) on the mainland also applies on the small islands (LBH Padang, 2005). Even though they are separate from the mainland, the treatment dealing with the utilization mechanism of land is the same. There is no difference between land on the mainland and land on the small islands. However, the distance between the community's settlement and the land is far enough, in fact some of them are quite far, and makes the lands on the small islands less managed and utilized or even not utilized at all.

The condition of the islands means there were no such utilization rights activities directly conducted by the local community. The distance between the islands and the settlement area

---

[10] According to the wisdom in the local community, the women are not allowed to get involved in the process of *Tulak Bala*. The women only contribute when the sacrificed animal (the goat) is prepared to cook. This cooking marks that the ritual is about to end.

[11] Angku Bujang Sabaleh is considered to be the super natural man by the local community of Nagari Sungai Pisang. He comes from Muko-muko, Bengkulu. According to the community, before Angku Bujang Sabaleh, it was Angku Sabaleh who led the *Tulak Bala* ritual.

[12] The light conversation was actually another way for the researcher to gain information about the rights that the community owned to utilize the islands. It was effective because through this technique, the informants were free to talk about anything without feeling like they were being interrogated, especially the women. The researcher called this type of interview "light conversation" because he spoke with them not as a researcher, but as a newcomer to the village.

[13] This occurred when the community of Nagari Sungai Pisang held the *Tulak Bala* ritual (a ritual to eliminate the bad fortune from the *nagari*). The researcher might join all the ritual activities, which are actually forbidden for women.

is the main reason for this situation. Besides, the high costs required to conduct such utilization activities makes the islands not receive priorities as cultivated land.

All of the utilization rights that the local community has on land on the mainland can be implemented also on the land in the small islands, as long as the rights are available. By the condition of Sikuai Island and Pasumpahan Island, which have been utilized by the investors, the utilization rights of the community cannot be implemented.

Even so, the local community still can take benefits from the small islands. It is local fishermen who usually visit the islands as a shelter from the storm or as stopping point to rest when they are fishing. Fish resources, land resources, plant resources, scenic resources, and water and ocean resources are the resources that are taken by the local community from the small islands. Plant resources are in the form of firewood and coconut fruits. The shorelines of the islands are also used by the fishermen as their base point by using a *pukek tapi*[14] fish catching tool. The scenic beauty of the small islands and their underwater views are sometimes used by the local community as an alternative for generating income. Local fishermen are allowed to catch the fish that surround Pasumpahan Island, but not those that surround Sikuai Island (Fig. 5.3).

Since the islands were leased to the investors (Sikuai Island in 1983 and Pasumpahan Island in 1994), the local community is no longer able to utilize the islands. The leasing agreement causes the local community to lose several rights in a certain period of time. By this agreement, the local community transferred their exclusion rights until the leasing period is over. Therefore, the rights for utilizing the islands are determined by the leaseholders as they are holding the exclusion rights and the rights to determine who will have an access right and how those rights may be transferred to others.

As a consequence of this agreement, the local community only has access rights to Sikuai Island; it is determined by the manager of the resort. No extraction activities by the local community are allowed on this island. Meanwhile, the tourism development or construction on the island involved the local community as the supplier of some construction materials and some as the workers (Fig. 5.4).

On Pasumpahan Island, the local community is the authorized user, as they have the access and withdrawal rights of the resources on the island. The leasing agreement causes the

FIG. 5.3   The fishermen extend net surrounding the shoreline of Pasumpahan Island.

[14] *Pukek tapi* is a wider net with rings. To catch the fish using this kind of fishing equipment requires a team, which usually consists of four to five people.

FIG. 5.4   Sikuai Island resort officially open.

reduction of rights. The local community can take some of the plant resources available on the island, but only for consumption. Sometimes, the community took some of the coconut fruits to be used for any event held by the *nagari*. They took the coconut fruits from the island only when the coconut fruits that grew on the mainland were not sufficient enough. The economic benefit from the island is only received by the community from payments of the island (IDR 2 million per year). Another plant resource that is being taken by the local community is firewood as cooking fuel. It was very beneficial for them to use free firewood rather than using kerosene that cost them money and was not always available due to shortages.

As the authorized user of both islands, the local people cannot complain about every policy created by the proprietor of Pasumpahan Island and the owner of Sikuai Island. Because as the authorized user, the local community cannot have the rights to devise their own harvesting rules.

Seeing the prevailing condition and situation of Sikuai Island and Pasumpahan Island from the property rights regime developed by Lynch and Harwell (2002),[15] there is no regime that precisely describes the property rights regime of Sikuai Island and Pasumpahan Island. But, the group-individual property rights regime is quite close to the property rights condition of these small islands.

These islands are sold and leased not in the name of the state,[16] but in the name of a group of people (the local community of Nagari Sungai Pisang). And these islands are leased to an individual (Pasumpahan Island) and a corporation (Sikuai Island).

[15] According to Lynch and Harwell (2002), there are four spectrums of property rights: private-individual, public-individual, public-group, and private-group.
[16] According to the state law, all of the small islands that are located within the jurisdiction of the Republic of Indonesia belong to the Republic of Indonesia. But in reality, the leasing agreement and the ownership certificate of Sikuai Island shows that there is an absence of state law. In the leasing agreement, the local community of Sungai Pisang is represented by the Penghulu of four clans who is the lessor of the island.

5.5 UTILIZATION RIGHTS IMPLEMENTATION

As Sikuai Island is becoming a tourism resort, the implementation of the utilization rights by the local community are also being affected. The manager of the resort issued certain regulations related to the implementation of the rights. The regulations stated that the local community could enter Sikuai Island only when there were no guests at the resort. But there is an exception. When there is a storm, not only the local fishermen, but also all the fishermen who surround the island, are permitted to enter the island for shelter. At Pasumpahan Island, there is no such regulation. The local community can enter the island and extract the resources of the island freely. But before they enter and extract the resources, they have to get a permit from the occupant of the island. It is just a matter of manner; respecting the people who are known to occupy the island.

Considering the implementation of the utilization rights at the customary land on the mainland, there is a certain mechanism that applied in Nagari Sungai Pisang. Based on the interview with the Penghulu of Koto Piliang clan, there is an unwritten mechanism for the people who want to utilize the land in Nagari Sungai Pisang. The people who want to utilize the customary land of the *nagari* must be domiciled at Sungai Pisang. Nevertheless, this does not close the opportunity for an outsider who is not domiciled at Sungai Pisang to utilize the land of Nagari Sungai Pisang (an example is the investor on Pasumpahan Island).

At first, the people who want to utilize customary land are classified into two groups. The first group is the people who have an identity card issued at Sungai Pisang, but not are affiliated with any of the clan in Sungai Pisang. The second group is the local people of Nagari Sungai Pisang and are members of one of the clans existing in Nagari Sungai Pisang. The people who do not have a Sungai Pisang ID card and are also not settling at Nagari Sungai Pisang were included in the first group. These groups obtained different rules of requirements. From this point, it can be seen that a legal pluralism occurred at Nagari Sungai Pisang. The implementation of state law (having a Sungai Pisang ID card) becomes the consideration for giving the license to utilize the customary land, which is determined by *adat* law. The incorporation of part of the state law into *adat* law implementation can be seen (Moores and Gordon, 1987 in Saptomo, 2004).

5.6 PROBLEMS DURING UTILIZATION RIGHT IMPLEMENTATION

In managing the customary land, known in *Minangkabau* culture as the principle of communalism, it does not mean that the practice of land management is free from problems and conflicts. The conflicts that might happen in customary land management in the *Minangkabau* community are mentioned below (Kanwil BPN Prop Sumbar, 2000 in LBH Padang, 2005).

a. Internal *Nagari* Conflict: conflict that happens in the *nagari* that involves the *anak nagari* of the related *nagari*. An example is the conflict of the unfairness of land distribution of the customary land.
b. Internal Clan Conflict: conflict that happens that involves the local people of a clan in a *nagari*.

c. Internal *Kaum* Conflict: conflict that happens in a *kaum* in a *nagari*. Usually this happens between one *paruik* with another *paruik* competing for the customary land of *kaum*.

d. Cross-*Nagari* Conflict: conflict that happens between one *nagari* and another *nagari*. Cross-*nagari* conflict usually happens in terms of determining the borders of the *nagari*.

e. Conflict with Outsider: conflict that happens between the community in a *nagari* and the people that are coming from outside of the *nagari*. This conflict can happen at the *nagari*, clan, or *kaum* level.

Conflicts or problems between the *nagari*, clans, or *kaum* almost never happen in Nagari Sungai Pisang; even with the outsiders who want to cultivate the customary land of the *nagari*. The local people of Nagari Sungai Pisang welcome outsiders. They are accepting everybody that comes to their area with good will.[17]

There were almost no problems during the implementation of the community's rights on both islands, especially at Sikuai Island. Complains and grievances of the local community were never delivered to the resort management of Sikuai Island (preconflict stage or, in other words, a monadic conflict)[18] (Saptomo, 2006). Actually, the local community had never used the litigation mechanism to solve a conflict, not once. The case was brought to trial by the disputing parties. The disputing parties at that time were the local community and the previous investor in Pasumpahan Island, because they claimed they owned the island. Misunderstanding between the local community and the proprietor of Pasumpahan Island occurs not too often. The local community preferred to choose negotiation to solve any problems that occurred.

When there was an event at the Nagari Sungai Pisang, the local people usually asked for the assistance of the foreigner at Pasumpahan Island. This assistance was sometimes in the form of resources (coconut fruits) and funds. But the assistance intended here is assistance for conducting an event; not assistance to support the livelihood of the local community of Nagari Sungai Pisang.

No such enormous conflict ever happened that involved the local people and the foreigner and his workers. Whatever problems existed were able to be solved harmoniously. The local community never let conflicts arise. Any small problem was solved immediately; it never waited till tomorrow. Relationships among members of the local community itself were close

---

[17] One example is the outsider who intended to open a clutch plantation on the customary land of the *nagari*. Before opening the areas for cutch plantation, they asked for a utilization license to the Penghulu of four clans in Nagari Sungai Pisang. Then the Penghulu of the clans held a village meeting attended by *ninik mamak, bundo kanduang* (the mother), *urang tuo* (respected man in the community), and Lurah. This meeting was held as a hearing meeting. The decision whether to accept or to reject the request of the outsider to open a clutch plantation on the customary land of the *nagari* was determined by the Penghulu of the clans based on the result of the meeting. But the common situation that prevailed regarding the permission license to utilize the land has always been given. As long as the "*Adaik diisi; limbago dituang*" is fulfilled, there is no problem in issuing the utilization license.

[18] This problem is called the preconflict stage because at this point, the in condition shown in the social relationship between Sikuai Resot management and the local community of Nagari Sungai Pisang, only one side feels disadvantaged by the other party so that condition makes the disadvantaged party have grievances; which is the local community of Nagari Sungai Pisang.

and tight. People never let the outsider know their problems. It is embarrassing if the outsiders know their problems, especially if it is a small problem. The community sees the conflict as a common thing in the society.[19]

5.7 PROBLEM-SOLVING MECHANISM

In the perspective of the lesson of *Minangkabau* culture, any conflicts (internal *nagari* conflict, internal clan conflict. internal *kaum* conflict. cross-*nagari* conflict, and conflict with an outsider) had to be solved by deliberation (*musyawarah mufakat*) to look for fair solutions.

When solving a conflict or problem, there are certain things to be paid attention to. A way must be found where nobody loses (all the conflicting parties were not being disadvantaged; a win–win solution). If we make a bad decision, we have to apologize to avoid creating another conflict in the future. The problem is solved, but nobody gets hurt. And the way to reach this point is by sitting together deliberating the main cause and talking with no emotion.

In solving conflicts, the *adat* law of *Minangkabau* placed the truth at the highest level with the *adat* leaders (Penghulu) as the executor of the truth, which is based on the deliberation. Even though the Penghulu in the lesson of *Minangkabau* culture has been given the authority to solve conflicts, they are still not allowed to act authoritatively, as it is said in the proverb *"mamancuang putuih, mangauih abih."* This proverb means that the Penghulu is not allowed to solve a conflict based on his will only. The Penghulu has to put forward the principles of deliberation. He must listen to all opinions regarding the conflict so that in the end a solution can be arrived at that is equal and satisfies the conflicting parties. The deliberation aims to reach a decision that is sincerely fair.

At Nagari Sungai Pisang, the condition is not much different. The Penghulu of four clans are the ones who used to solve the problem in the *nagari*. But sometimes their position when solving a problem is not the same as when they are solving other problems. Besides being the Penghulu, they are also the *mamak* of their niece and nephew. There is a hierarchy problem that later determines who will be the problem solver. This hierarchy is based on the type of conflicts that usually happened in the *Minangkabau* community in terms of customary land management.

The first type is called internal *nagari* conflict (problem). This problem is usually being solved by the Penghulu of all clans in the *nagari*. Any problem related to the welfare of the whole community of the *nagari* is also solved by the Penghulu of all clans; in the Nagari Sungai Pisang context it solved by the Penghulu of four clans. A problem involves the members of a *kaum*, or it is called the internal *kaum* conflict, then the Penghulu of the *kaum* has the obligation to solve the problem of his *kaum*'s member. And so if the problem (internal clan conflict/problem) involved a member of a clan, then the clan's Penghulu is needed to solve the problem.

---

[19] This impression was gained during the interview and light conversations with the key informants and the local people. When the researcher asked about the conflicts or problems that have ever happened between the local people and the foreigner, they often did not directly answer the questions. They described the conflicts by saying *"namonyo urang iduik badampiangan tu ado tasingguang agak saketek. Kok ta lantuang ka naiak, ta lendo ka turun; itu biaso. Ndak manjadi masalah bana dek awak…"* (When we are living side by side, there must be a clash between us. It is not a big matter for us.)

When the problem involved two or more *nagari*, if the Penghulu of the clans of each *nagari* cannot solve the problem, usually they agree to bring the problem to be solved by the KAN. But such a condition rarely happened in Nagari Sungai Pisang. In solving the *nagari* matter, the Penghulu can ask for an opinion from another Penghulu, even a Penghulu from a different *nagari*. This showed that in making a decision to solve a problem, every consideration and view are important things, especially when related to the welfare of the community or related to the cooperation relationship of the *nagari* and the other party. When facing a problem with the resort management of Sikuai Island, the local community just left the problem alone or lumped it (Nader, Laura, and Harry F. Todd, Jr., 1978 in Saptomo, 2006). Lumping the problem was one of the problem-solving mechanisms that had been stated by Nader (1979). Such a mechanism can be found in this community.

There is an article considering the dispute resolution mechanism in the leasing agreement with the investor of Pasumpahan Island. The article stated that whenever there is conflict or a problem between the contracting parties, and then deliberation will be used as the mechanism of problem solving. The involvement of Lurah in solving a problem in Nagari Sungai Pisang often happened. In any decision regarding the interest of the *nagari*, Lurah is always being asked for his opinions. Because according to the Penghulu of four clans, every decision made, however, can touch the state law. So the involvement of Lurah is necessary as he is the person who best understood the state law. There was no such strict mechanism for solving a problem in Nagari Sungai Pisang. The conflicting parties can solve the problems by themselves directly. For a more complicated problem, the Penghulu or *ninik mamak* or *urang tuo* is required to solve it. But such types of problems seldom happen in the *nagari*, in terms of resource utilization and rights implementation on both Pasumpahan Island and Sikuai Island. Before they make any decisions, they discuss the goods and the bad possibilities for making a certain decision. Therefore, deliberation (*musyawarah mufakat*) needs to be reached in any meetings they hold. Any objections and disagreements will be delivered in the meetings. The objections and disagreements in the end are useless.

5.8 CONCLUSIONS

Before the islands were leased, the local community had a complete bundle of rights, as the owner. But since then, the local community only has the access rights to Sikuai Island and access rights to and withdrawal rights on Pasumpahan Island. Since the Sikuai Island became the tourism resort, the manager of the resort issued certain regulations that affected the implementation of the local community's rights on the island. A certain condition prevailed to implement the rights. In reverse, at Pasumpahan Island there were no such regulations. Almost all the time whenever they want to, the local community can enter and extract the resources from the island, as they are holding the access and withdrawal rights as the authorized users.

The problems that occurred during rights implementation on both islands were not too significant. Misunderstanding between the local community and the proprietor of Pasumpahan Island occurs not too often. Complaints and grievances of the local community were never delivered to the resort management of Sikuai Island (preconflict stage).

In solving most of the problems that occurred, the local community chooses alternative dispute resolution mechanism. For certain big problems, local community even brought the case to trial once. Aside from that case, when big problems occur next, the local community prefers to lump it; considering the lack of money, knowledge, and human resources. The involvement of Lurah in the problem-solving mechanism also contributes to solve problems.

References

Annan, K., 2004. Report of the Secretary-General, "Review of progress in the implementation of the programme of action for the sustainable development of small Islands developing states". United Nations Department of Public Information–DP/I/2348F/Rev.1—November 2004.

Anomymous, 2007. Monografi Kelurahan Sungai Pisang. Padang.

Central Bureau Statistic (BPS), 2006. Bungus Teluk Kabung in Figures. Padang.

Creswell, J.W., 2003. Research Design, Quantitative & Qualitative Approaches. KIK Press, Jakarta.

Dahuri, R., 2000. Pendayagunaan Sumberdaya Kelautan untuk Kesejahteraan Rakyat (Kumpulan Pemikiran Rokhmin Dahuri). In: Budiman, A., Iakwati, Y., Widyanto, U. (Eds.), Lembaga Informasi dan Studi Pembangunan, Jakarta, Indonesia.

Gaskell, G., 2000. Individual and group interviewing. In: Bauer, M.W., Gaskell, G. (Eds.), Qualitative Researching With Text, Image, and Sound: A Practical Handbook. SAGE Publications, London.

Khaka, E., 1998. Small Islands, Big Problems: Outlines the Special Water Needs of Small Islands, Our Planet 9.4—March 1998. UNEP.

Lynch, O.J., 1999. Promoting legal recognition of community-based property rights, including the commons: some theoretical considerations. In: Paper Presented at a Symposium of the International Association for the Study of Common Property and the Workshop in Political Theory and Policy Analysis, Indiana University, Bloomington, Indiana, June 7.

Lynch, O.J., Chaudhry, S., 2002. Community-based property rights: a concept note. In: A Center for International Environmental Law Issue (CIEL) Brief for the World Summit on Sustainable Development 26th August–4th September.

Lynch, O.J., Harwell, E.E., 2002. Whose Resources? Whose Common Good? Towards a New Paradigm on Environmental Justice and the National Interest in Indonesia. Center for International Environment Law (CIEL) in collaboration with Perkumpulan untuk Pembaruan Hukum Berbasis Masyarakat dan Ekologis (HuMa), Lembaga Studi dan Advokasi Masyarakat (ELSAM), Indonesia Center for Environmental Law (ICEL), International Centre for Research in Agroforestry (ICRAF), Jakarta.

LBH Padang, 2005. Kearifan Lokal dalam Pengelolaan SDA (Kekayaan Nagari Menatap Masa Depan). INSISTPress.

Mook, E., 1998. GEF: Helping Small Islands Developing States. Complementary Article in other issues: Elizabeth Khaka: Small Islands Big Problems (Freshwater). http://www.ourplanet.com/imgversn/103/13_gef.htm.

Nader, L., 1979. Disputing without the Force of Law. Yale Law Journal 88, 998.

Rahmadi, T., 2004. Law and policy regarding Natural Resource Management with special reference to West Sumatra and the role of Customary Law in the Natural Resource Management with special reference to West Sumatra. In: A paper presented in Proceeding International Seminar on Integrated Natural Resources Management with Special Reference to Sumatra and Workshop for Master's Degree Curriculum Development. Padang, 25–27 January 2004, Indonesia.

Saptomo, S., 2004. Local potential in land control and natural resource use. In: Paper Presented at International Conference on Land and Resource Tenure in Changing Indonesia "Questioning the Answer", Hotel Santika Jakarta, Indonesia, 11 October.

Saptomo, A., 2006. Pengelolaan Konflik Sumber Daya Alam Antar Pemerintah Daerah dan Implikasi Hukumnya, Studi Kasus Konflik Sumber Daya Air Sungai Tanang, Sumatera Barat. J. Ilmu Hukum 9 (2), 130–144.

Scialabba, N. (Ed.), 1998. Integrated Coastal Area Management and Agriculture, Forestry and Fisheries. FAO Guidelines. Environment and Natural Resources Service. FAO, Rome.

Shepherd, G., 2006. As reported at "Workshop on the ecosystem approach and customary practice in protected areas in small Islands", 12–16 December 2006, Bangkok, Thailand.

Shivakoti, G., 2006. Lecture Note on Natural Resources Management and Planning and Policy, at Integrated Natural Resource Management Study Program. Post Graduate Andalas University, West Sumatra, Indonesia.

Stainback, S., Stainback, W., 1988. Understanding and Conducting Qualitative Research. Kendall/Hunt Publishing Company, Dubuque, IA.

Suhana, 2005. Benarkah Pulau-pulau Perbatasan Terancam Hilang? Sinar Harapan, 30 Maret, 2005, http://www.sinarharapan.co.id/opini/index.html.

UNEP Islands Web Site, 2004. Small Islands Environmental Management. http://islands.unep.ch/siem.htm.

Gender Inequality of a Fishing Family in a Small-Scale Fishery: A Case Study on a Fishing Family in Korong Pasir Baru, Nagari Pilubang, Sungai Limau Subdistrict, Padang Pariaman District, West Sumatra, Indonesia

Nofriyanti

Regional Body for Planning and Development, District of Padang Pariaman, Indonesia

6.1 INTRODUCTION

Fisheries are a source of income for over 100 million people. The majority are employed in small-scale fisheries in the developing world; 90% are from Africa and Asia, where poverty among coastal and rural communities is often high. The fishing families usually live in very poor conditions without any basic amenities. This is often exacerbated by their remote locations. In Indonesia, coastal fishers in Java are poorer than rice farmers and freshwater fish farmers. They often have very poor living conditions without the basic amenities. Similar to Malaysia, social mobility is low and children of the fishing family normally follow their parents' footsteps, and those from very poor families are forced to work before they attain the legal age to enter the workforce. Dwindling catches, however, have forced some of these children to seek jobs outside the fishing sector. It has been observed that women workers involved in the rice fields look for jobs in the fisheries-related sector when planting paddy became mechanized and new technology was introduced. A study carried out by Susilowati (1991) showed that in general there are about 1.5 million fishermen who live in the fishery area in Indonesia and about 60% of fisher families live below the poverty line. Furthermore, Manurung (1983) emphasizes that fishermen in Indonesia have very low incomes and they are the poorest of the poor. The fishermen families living in coastal areas are known as the most underdeveloped families group (Zein, 2004). The traditional fishermen families are characterized by a low level of production assets, usually catching fish without a motorized boat. Low production assets also affect their access competence toward production of fishing ground, which results in too low production and a low level of income.

Padang Pariaman District is one of eight fisheries areas in West Sumatra, with a 60.5km coastline. Fishing is the main livelihood activity of the local people in Padang Pariaman. Six of seventeen districts in Padang Pariaman are located in fishing areas and the number of fishermen was noted as about 10,300 individuals. Facts indicated that persistent poverty and deteriorating economic conditions have forced many women from poor rural households to work outside their homes and venture into varied economic activities while at the same time continuing to perform their traditional household duties. The burden of crises often falls on the shoulders of women of fishery households, forcing them to put in longer working days and perform a wider range of income-generating activities in harmful working conditions. The low standard of living and reduced fishing catches due to decreased fish stocks, means that women are increasingly engaged in productive activities to supply the family with its daily needs. Women are playing an important role not only in domestic-based activities, but also in productive activities to support their family's economy.

The important involvement of women in natural resource-based livelihoods and resource management in the developing world has long been acknowledged, but rarely been valued equally with the contribution of men (GWA, 2006). The role of women is often unconsidered in its economic and cultural aspects, so that it will cause gender inequality. Unequal relationships between women and men will affect their opportunities to participate in and benefit from development interventions. For that reason, this study focuses on gender equality within the fishing family and is aimed to answer why and how gender inequality occurs in the fishing family. This study will also analyze factors that may contribute to the work distribution and time allocation between fishermen and women.

6.2 RESEARCH OBJECTIVES

The objectives of this study are

1. To explore roles and responsibilities of men and women in the fishing family in terms of productive, reproductive, and social activities and describe their time allocation.
2. To assess the genderwise access to and control of the family assets, resources, and decision-making process in the fishery area.
3. To identify factors that may influence gender inequality in the fishing family.

6.3 METHODOLOGY

6.3.1 Study Sites

This study was conducted in a small-scale fishery in Korong Pasir Baru, Nagari Pilubang, Sungai Limau Subdistrict, Padang Pariaman District. Padang Pariaman District is an appropriate site location for this research as in Padang Pariaman fishing is the main livelihood activity of the communities living in this area. Sungai Limau Subdistrict is selected for the following reasons: first, Sungai Limau is the center of coastal areas in Padang Pariaman District; second, production of marine fish in Sungai Limau are the highest in Padang Pariaman District; and third, the number of sea fishermen and fisherwomen in Sungai Limau Subdistrict are higher than the other subdistrict. Based on data in Nagari Pilubang, in Korong Pasir Baru there are 304 households (comprising 892 men and 955 women) and about 220 (approximately 73%) of all households are fishermen.

6.3.2 Research Design

The observed facts and situations related to gender inequality in the fishing family are presented to provide an important insight into a given situation. Usually most of the studies in the context of West Sumatra have only explored women's contribution in supporting household income. The present study is aimed at providing more exploration of male and female roles and gender inequality. Hence, it can explain the observed facts and situation relative to construction of theories or the formulation of a hypothesis for further research.

6.3.3 Data Collection

Primary data for the present research was obtained by (a) field reconnaissance surveys; (b) participatory observation; (c) in-depth interviews, by interviewing relevant key people such as the fishing committee, village and *korong* head, women leaders in the group, an *adat* figure, an elder person in the community, and so on; and (d) household surveys. On the other hand, the secondary data are in the form of government documents, project reports, maps, and reports on previous relevant studies on the research topic. The sources of secondary data are government offices, the *nagari* office, Central Bureau of Statistics, Ocean and Fishery Department, NGOs, Graduate Program Library at Andalas University, PSI-SDALP Library at Andalas University, and also electronic data derived from Internet sources.

6.4 FINDINGS AND DISCUSSION

6.4.1 Gender Roles and Responsibilities in a Fishing Family in Pasir Baru

One way to see the roles of men and women in such activities is to look into the time they spend in such activities.

From Table 6.1 it can be seen that the total time allocation of men and women on three activities is similar. Women spend on average 14.46 h/day whereas men spend on average 15.76 h/day in such activities. The difference is that women spend much of their time on reproductive work and are busy with their domestic tasks while men spend much of their time in productive activities to earn income. The rest time and other time for men and women is also quite similar. Women can enjoy their rest time for 9.54 h/day and men can enjoy their rest time for 8.24 h/day.

There are many differences of time allocation on reproductive activities between a woman who has productive work and a woman who has no productive work. Comparing the time

TABLE 6.1   Time Allocation of Men and Women on Different Activities

Type of Activity	Men (h/day) N=30	Women (h/day) N=30
Reproductive	1.04	6.39
Productive	14.47	7.91
Social	0.25	0.16
Total of time allocation	15.76	14.46
Rest time and others	8.24	9.54
Total	24.00	24.00

N, number of respondents.

allocation for reproductive activities between a woman who has economic activity with a woman who does not have economic activity indicates that a woman who has no economic activity spends more of time on reproductive activities than a woman who has economic activity (Table 6.2).

The positive impact of women involved in productive activities, even though they spend less of their time on reproductive activities, is that they are able to contribute to the family

TABLE 6.2   Time Allocation of Women in a Fishing Family

Type of Activity	Women Who Have Productive Activities (h/day) N=27	Women Who Don't Have Productive Activities (h/day) N=3
Reproductive work: cooking	1.13	2.67
Washing dishes	0.25	0.33
Washing clothes	1.13	1.67
Fetching water	0.22	0.42
Firewood collection	0.16	0.57
Cleaning house	0.24	0.33
Caring for children	2.13	3.33
Ironing	0.22	0.33
Total of time allocation in reproductive work	5.48	9.65
Productive work	7.91	0
Total of time allocation in reproductive and productive work	13.39	9.65
Rest time and others	10.61	14.35
Total	24	24

N, number of respondents.

income and help their husband in fulfilling the family's daily needs. They can contribute to family income by doing take-home embroidery, running a small home-based grocery store, fish drying, and fish processing. They can do these activities in their own house and they can still supervise their children.

6.4.2 Access and Control of Men and Women on Family Assets and Resources

Access is the right and opportunity to obtain, make use of, or take advantage of something, whereas control is the authority or ability to manage something or to exercise authoritative or dominating influence over something (Vernooy, 2006). Tisch (1992) as cited in Aryal (2002) states that one of the several key issues that need to be considered in gender concerns is access to and control of resources and benefits that contribute to family welfare. Access is the freedom or permission to use a resource, whereas control is the power to decide whether and how a resource is used.

Gender inequality can be seen in various forms. One of them is in terms of access to and control of resources and the decision-making process. Women's relative lack of control over resources is a key factor perpetuating gender inequality. In reality, access and control are related to each other. Access to and control over family assets and resources is a mirror of one's economic status within the household, which ultimately determines economic positioning in the household and community (Adhikari, 2001). In the study area the level of access between men and women to their family assets is similar. Both have the same opportunities and chances of making use and taking advantage of their family assets. However, there is a higher influence of men in terms of access and control over resources that exist in their area. It happens in deciding many things related to fishing activities, such as deciding to catch the fish/crabs and other marine production from the sea, deciding to sell the fish from fishing, deciding the price of fish, using technology for fishing, borrowing money from Koperasi to run fishing and nonfishing activities, and deciding about purchasing and selling/mortgaging the family assets is also under the men's control. Men are considered as the right person to purchase/sell high-valued or expensive things, whereas women do not have enough courage to carry out such activities. Overall, the level of access and control over resources of men in the study area is much higher than women. It can be concluded that inequality has occurred in terms of access and control over resources between men and women in the study area.

6.4.3 Factors Influencing Gender Equality

Various factors may influence gender equality. These factors include family characteristics, sociocultural attitudes, educational level/general knowledge, religious beliefs/practices, and formal legal system/policy (adopted from ADB, 2002, "Gender checklist—agriculture").

Of all the issues that influence society, none is more profound than gender: the countless, unspoken cultural rules that differently govern the behavior of females and males in every country in the world, from the day they are born. The difference between men and women show up clearly in the division of responsibilities at home and in their communities. In the fishing sector, the situation is no different. Fishing is considered as men's work. All of the activities related to the fishery are men's activities while women are considered as housewives who have responsibilities in domestic tasks and caring of their children and family.

Among the various factors influencing the role, position, opportunity, chance, and division of work between men and women in this area, sociocultural factors and religious practices have high influence on gender equality in the study area. The cultural, custom, values, and *adat* that have existed in this area for a long time can't be changed easily. It is in accordance with the Development Assistance Committee (1998) statement that culture influences access to and control of men and women over resources, and participation in decision making. Culture is sometimes interpreted narrowly as "custom" or "tradition," and assumed to be natural and unchangeable. Despite these assumptions, culture is fluid and enduring.

A study carried out by Gurung (2006) about women in fisheries in Nepal found that it is usually said that a home usually means more to women than to men. The male members of the house are totally disinterested in doing domestic work. Women usually become frustrated and tired as they have to do almost everything in and around the house. Men usually have the idea that all household work is women's responsibility. Even though the domestic work is vital, it is invisible, unpaid, undervalued, and unrecognized. As a result of a social emphasis on the role of women as domestic workers, they are more likely to miss out on opportunities for training, education, and information. So women's chances for progress lags behind those of their male counterparts. There is an unequal opportunity for women in education, training, and leadership biased by traditional values and stereotypes that domestic work is only women's responsibility. A further study carried out by Chando (2002) about gender roles in a fishery in Tanzania also found that traditional beliefs and demographic factors have been found to be among the factors that hinder women's participation in public activities in many societies. In particular the culture of the people influences the division of labor in many communities, which in most cases classifies men as leaders and decision makers.

Gender roles are associated with social norms and perceptions that shape men and women's behavior differently. In Korong Pasir Baru society it is perceived that men should do the heavy work, know how to lead the family, and have broad knowledge. When a man becomes a husband, he should know how to earn income to feed family members. Men should know fishing activities, while women should know how to do household work. The position and roles of women in its social community tend to place women as the weak party in the community.

There were various forms of sociocultural constraints, such as female seclusion practice and the social perception that fishing is only men's activity. In many cases, women depended on their husbands, whom they considered as the family keeper. Usually men hope that their wives stay at home and do not go out of the home. Women need to get permission to be involved in productive activities where these are located out of the home. Men prefer their wives to do productive activities in their house, such as take in home embroidery, run a small in-house grocery store, sell the fish product, and so on. Those activities do not force the women to go out of the home. They can do productive activities and earn income without going out of the home and still be able to help their family members and finish their domestic work.

Beside sociocultural factors and religious practices, family characteristics and education are also important factors that may influence gender roles both in the family and in society. For example, educational level and family class are both factors that can influence the self-confidence of men and women. Respondents from the rich and educated class tend to have higher self-confidence compared to respondents from the poor and uneducated class. This problem can affect their role in being involved in social activities and also can affect their access or control over something.

6.5 CONCLUSIONS AND RECOMMENDATIONS

This study analyzes the gender inequality in the fishing family, which highlights the gender division of work, their time allocation in such activities, and also assesses their access and control over family assets and resources in the study area. Moreover, this study also explores the factors that may influence gender roles and gender equality in the fishing family.

The findings revealed that there is no inequality in terms of time spent by men and women on reproductive, productive, and social activities together. Moreover this study also found that the level of access of men and women is equal in terms of family assets. However, the access to and control over resources is biased in favor of men. Overall, the level of access and control of men over resources is much higher than that of women. It is also reflected in deciding many things related to fishing activities such as decisions about purchasing and selling of the family assets. However, gender inequality does not occur in this area because there is still equal enjoyment by men and women in access to their family assets and equality in the division of work between them.

Many factors may influence the role, position, opportunity, access/control between men and women in this area: family characteristics, sociocultural factors, religious factors, educational factors, and the formal legal system/institutional policy. This study found that sociocultural and religious beliefs have the highest influence on gender equality in the study area compared to other factors.

Based on the findings of the study, the following recommendations are forwarded:

- Developing opportunities for more income-generating activities and access to credit for women (wives of fishermen) in study area.
- Encouraging women to be active in community activities, the decision-making process, and involving women in such training held by the government or NGOs.
- Empowerment of group members.

The empowerment of group members has an effect on the community as a whole. By women's group membership, they gain confidence and respect. Women can now see and judge the outer world, become mobile, and do not just stay at home. The most noticed beneficial part of women in group involvement is that they can speak and express their opinion, which was suppressed for a long time due to traditional cultural beliefs that women should not be outspoken and should not complain at all. Education empowers women to gain more confidence, understand the complexity of management procedures, and manage things in a better way. Generally, women are totally dependent on their husband as they usually do not have any other skills except embroidery and drying fish. It is better to launch training programs on such productive activities. When they have more skills they can earn more money, develop bargaining power, and can increase their self-confidence. Training on new processing and preserving techniques, making fish pickle and/or fish pasta, which have great demand in the market, can generate more income for the fishing families. These activities can be carried out in their own houses with less effort and the additional income generated through these activities will motivate other family members to provide helping hands.

Acknowledgments

I am most grateful to my first supervisor, Dr. Ir. Rudi Febriamansyah, MSc, for his unique professional supervision and meticulous comments, which have been very valuable throughout this work. I am obliged to say without his patient guidance and encouragement this work could not have become a reality. Then, to my second supervisor, Dra. Ranny Emillia, M.Phil, for her valuable comments and suggestions. My deep gratitude to Andalas University and the Asian Institute of Technology (AIT) Bangkok for awarding me a fellowship and scholarship, respectively, to pursue my masters degree at Andalas University. I am also grateful to the Center for Irrigation, Water-Land Resources and Development (PSI-SDALP) at Andalas University for providing me valuable information and managing all of my scholarship during my study at Andalas University.

References

Adhikari, L., 2001. Women's Participation and Empowerment in Nepal: An Evaluation of Rural Urban Partnership Program. School of Environment, Resources and Development, Asian Institute of Technology, Bangkok.

Aryal, S., 2002. Rural Women's Participation in Agricultural Activities and Decision Making in Household; a Case Study of Raviopi Village in Nepal. Asian Institute of Technology, School of Environment, Resources and Development, Bangkok.

Asian Development Bank (ADB), 2002. Gender checklist—agriculture, Biro Pemerintahan Nagari/Kelurahan Setda Prop, Sumatra Barat. 2006, Data Isian Monografi Nagari Pilubang Kecamatan Sungai Limau Kabupaten Padang Pariaman, Propinsi Sumatra Barat.

Chando, C.M., 2002. Gender Roles in Fishery Planning and Projects. The Case Study of Coast Region in Tanzania. Department of Social Science and Marketing, Norwegian College of Fishery Science, University of Tromso, Tromso.

Development Assistance Committee (DAC), 1998. Evolution of the thinking and approaches on equality issues. DAC Source book on Concepts and Approaches Linked to Gender, OECD, Paris.

Gender and Water Alliance (GWA), 2006. Gender and Fisheries. www.genderandwateralliance.org.

Gurung, M., 2006. Women in Fisheries: A Study on Women's Coping Strategies in Fishing Communities in Lake Begnas, Nepal. Norwegian College of Fishery Science, University of Tromso, Tromso.

Manurung, T., 1983. Suatu Tinjauan Kriteria Nelayan Kecil dan Masalah Pembinaannya di Jawa. Badan Penelitian dan Pengembangan Pertanian, Departemen Pertanian, Jakarta.

Susilowati, 1991. Kontribusi Keluarga Wanita Nelayan dalam Pengembangan Usa HAPerikanan di Indonesia. InstalasiPenelitian dan Pengkajian Teknologi Pertanian Bogor, Jawa Barat.

Vernooy, R., 2006. Social and Gender Analysis in Natural Resource Management; Learning Studies and Lesson From Asia. International Development Research Centre, Ottawa, ON.

Zein, A., 2004. The Role of Fisher-Women on Food Security at the Traditional Fishermen Household of West Sumatra, Indonesia. Bung Hatta University Press, Padang.

Women's Participation in a Rural Water Supply and Sanitation Project: A Case Study in Jorong Kampung Baru, Nagari Gantung Ciri, Kubung Subdistrict, Solok, West Sumatra, Indonesia

Yuerlita

Andalas University, Padang, Indonesia

7.1 INTRODUCTION

7.1.1 Background

One of the main problems faced by developing countries is providing safe drinking water for both urban and rural communities. In the context of the rural community, the problem is commonly caused by the scarcity of water sources. It is also difficult to access water resources because of its geographical condition. Moreover, internal factors, such as an increase in population pressure and conflict, privatization, and changing tenure arrangements, poverty, social differentiation, and environmental degradation have affected the type, quality, accessibility, and reliability of water resources.[1]

To enhance the sustainability of the rural water supply scheme, many projects, whether facilitated by government or nongovernmental organizations (NGOs), have applied a participatory approach to the project cycles, including planning, implementation, operation

---

[1] In Mwisi, Uganda, population growth has forced people to settle in ecologically sensitive areas such as hilltops and wetlands. This will then make them lose previously available water resources and hard for them to find water resources in their new settlement (Thompson et al., 2002).

and maintenance, and monitoring and evaluation processes. The participatory approach is also reflected after project completion. The management and maintenance of the water supply schemes is handed over to the community as the beneficiaries. Water supply schemes with a participatory approach should give better sustainability of Rural Water Supply and Sanitation (RWSS). Participatory evaluation has shown that past water supply and sanitation intervention supported by aid agencies in Indonesia poorly addressed the equity of access for the poor, especially women. One of the important water users is women, who play a central part in the provision, management, and safeguarding of water. Women are responsible to provide safe and adequate water for family needs as well as for managing water to fulfill water needs appropriately. On the other hand, their high contribution to the family's livelihood is not matched by an equal access to water. Millions of women in the world spend 1–6 h a day fetching water for livestock, home gardens, and domestic uses such as bathing, cooking, washing clothes, and so on.[2] Hence, it is important to study women's participation with an emphasis on equity between men and women related to their participation in project cycles for ensuring the sustainability of RWSS. The objectives of this study are to describe the history and performance of RWSS in the study area, to describe equal participation of men and women in the RWSS project, and to identify factors that constrain women to participate in the RWSS project.

7.1.2 Research Problem

Waterborne disease and diarrhea are some kinds of diseases caused by lack of access to clean water and improper sanitation facilities. This problem is mainly faced by developing countries including Indonesia. More than 31% of rural communities in Indonesia do not have access to clean water and more than 62% of the communities do not get access to sanitation (WDR 2005 in Robinson, 2005). The government, supported by national and international agencies, NGOs, and other institutions, executed many projects with the aim to provide clean, safe drinking water and better sanitation facilities. Many efforts related to the provision of clean drinking water and sanitation have been implemented by the Indonesian government. However, the performance of the water supply and sanitation sector in Indonesia is estimated to be lower compared to other countries in Southeast Asia (WHO-UNICEF, 2004, cited in a proposal of the National Program for Community Water Supply and Sanitation Services). A study by Sara and Katz (1997) on rural water supply sustainability in relation to project rules in some projects in some countries, including Indonesia, reported that Indonesia has the lowest score in physical conditions, operation and maintenance, consumer satisfaction, financial management, and willingness to sustain the system. Some of the problems are low quality of construction, experiencing a decrease in water flow, and lack of people's knowledge about coping with destruction. Some women as the main beneficiaries tend to use old water sources because they are not satisfied with the new water facilities. As a consequence, community's (particularly women) do not have a sense of ownership in the system, which causes unwillingness to sustain the system.

---

[2] http://www.worldbank.org/transport/rural_tr/imt_docs/ntk6c.pdf.

The above problems cause lack of project sustainability. The problems might be caused by many factors such as technical problems, institutional issues, and an unclear legal framework that guides project implementation. However, the level of community participation in all phases of the project plays an important role for the project outcomes. In this context, community participation means involving women as well as men. This is related to women's role in the family as users as well as water managers. Therefore, this study focuses on women's participation in the RWSS project, in the context of equal participation between men and women. Because many factors influence women's participation, the study also explores other factors including sociocultural, religious beliefs and practices, institutional arrangements, and legal frameworks.

Addressing women's strategic issues and ensuring the sustainability of project benefits is interrelated (Regmi and Fawcett, 2001b). Gender equality is the major constraint in a RWSS project that affects the sustainability of the RWSS scheme. This study investigates how women participate in the project, the gender equity practice of community participation in a RWSS project, and the constraints for women's participation. In this regard, there are three main questions to be answered in this study:

a. What are the conditions of the water supply and sanitation in the study area?
b. How do men and women participate in the RWSS project?
c. What factors constrain women from participating in the RWSS project?

7.2 METHODOLOGY

7.2.1 Research Design

This is an exploratory study with an appropriate research design given there are few studies related to this research topic, especially in the context of West Sumatran society. The study area is located in Jorong[3] Kampung Baru, Nagari[4] Gantung Ciri, Kubung Subdistrict, Solok Municipality, West Sumatra Province. Jorong Kampung Baru was selected as Jorong Kampung Baru was under a RWSS program called Water Supply and Sanitation for Low Income Communities (WSSLIC). Categorized as a success and a sustainable project based on the project report and interview with the project staff and researcher observations, the WSSLIC project applied a participatory approach in the project cycle. After the construction, the RWSS schemes were handed over to the local community for management.

The primary data are collected through field surveys. Data collection techniques applied in this study are field reconnaissance surveys, in-depth interviews with relevant key people, structured focus group discussions with women's and men's groups separately that were used to collect information on community participation with an emphasis on women's participation, household surveys conducted to assess community participation. and factors affecting their participation at individual levels.

---

[3] *Jorong* also called *korong* or *kampung*; hamlet comprising Nagari.
[4] A territorial unit in *Minangkabau* society, a *nagari* is a traditional, socioeconomic, and political body (Osmet, 1991).

The sample size of the household survey was determined as 30 households with 60 respondents. This is 14% of the total household number (216 households). One man and one woman were interviewed from each household. Random sampling was applied in this study. According to the newest data from *Jorong* headman, there are 13 high-class, 76 middle-class, and 127 low-class households. The required secondary data for the present research study is collected from government documents, project reports, maps, and reports on previous relevant studies to the research topic. The sources of those data are government offices, the *nagari* office, project office, the library, and electronic data collected from Internet sources.

7.3 RESULTS AND DISCUSSION

7.3.1 The Condition of the Water Supply and Sanitation Before the Project

Water sources in Jorong Kampung Baru are available in abundance, including spring, river, stream, unprotected well, and others. However, the location of springs are commonly far from the settlements. As a result, people cannot use springs as a healthy water source for drinking. There are three springs that are usually used by the community that are located in one area. They are about 2 km from the settlement. The debit of the spring is two lit/sec.

Most of the people in Jorong Kampung Baru do not have appropriate sanitation in their house. They usually defecate in a stream or pond near their house and also in the dry agricultural lands (*ladang*). This is because there are no water sources they can use to make a latrine in their house. The spring is quite far away from the settlement. People are not used to defecating using a latrine as they think it is dirty and improper.

Both bad latrines and the lack of a freshwater supply in this area caused diseases such as diarrhea and skin diseases, especially during the dry season. In fact, most of the community realized that what they have done is something unhealthy and not good for their health. However, the lack of facilities force them to continue these bad habits for many years.

7.3.2 The Condition of the Water Supply and Sanitation After the Project

There are 14 tap stands in Jorong Kampung Baru (out of 19 tap stands planned) and 14 groups of water users. There are four public latrines available. Public facilities, such as schools, a mosque, and a medical center are improved with provision for water supply and latrines. Almost all the tap stands are located by the roadside. Some of the households do not collect water from the tap because they already have a piped line connected inside the house. Other households have to walk 20–50 m to fetch water from the taps. They usually fetch water for drinking, cooking, and bathing. Other activities such as washing clothes and dishes are usually done at the tap stands. People usually fetch water in the morning; however, there is no particular time for this activity. They can collect water anytime they need it.

Although some men and children also fetch water, this is mostly considered a woman's job. Early in the morning, the women fetch water for drinking and food. After finishing all the household activities, they usually work in the fields to help their male counterpart and return back home late in the afternoon. Then they have to fetch water to prepare food for dinner. Women's workloads are decreased by the existence of the tap stand near their house. However, the existence of a tap stand does not change a woman's role as the main water collector.

The location of tap stands made the women feel embarrassed and reluctant to wash their clothes. They feel ashamed of being seen by men. They often have to wait until dark or carry out such activities early in the morning. Most of the women confess that they never think about the location of tap stands at the edge of the road affecting their activities. At first, women think that water facilities are built only for drinking water. In fact, now the water facilities are also for other needs. This is caused by the lack of information about the RWSS conducted in the area, as women rarely attend meetings and think that the technical arrangements are men's responsibility.

7.3.3 Women's Participation in Project Initiation and the Decision-Making Process

Sociocultural factors play an important role in the decision-making process. Related to its sociocultural system, the people in Kampung Baru usually discuss problems for the improvement of their *kampung*. This activity is known as *musyawarah*. However, this usually involves only men. Women rarely take part in *musyawarah* because the community thinks that women are only responsible for taking care of children and household activities, such as cooking, washing, cleaning the house, and sometimes helping their husband in *ladang*. Moreover, the meeting is usually held in the evening and that makes women reluctant to attend the meeting.

Table 7.1 shows the participation of men and women in the meeting. Eighty percent of men from 30 respondents attended the meetings, but only 63.33% of women from 30 respondents attended the meeting. Although more than 50% of the women attended the meeting, the frequency of their attendance is still less than men (see Fig. 7.1).

Only 4% of women attended the meeting more than three times, which is the highest frequency, while 36.67% of men attended the meeting more than three times, and 36.67% of the women never attended the meeting. The women only attend the meeting two or three times. The absence of women in a meeting is not merely caused by the objection of men to their involvement or hesitation and fear (Prokopy, 2004). In this study, women mainly claimed that their absence from the meeting is due to their roles as wives and mothers.

Women are not allowed to speak in front of older people, even older women (Prokopy, 2004). This situation is not the case in Kampung Baru. Women are no longer forbidden to speak in front of older people as long as they behave politely. This fact is also supported by informal discussion with the *ninik mamak*, who agree that there is no limitation for women to speak or give ideas in public discussion. However, a different situation is found in *adat* ceremonies, such as the meeting held by *ninik mamak* and other older respected people. Women

TABLE 7.1   The Tension of Attending a Meeting

Problem	Responses	Male (N = 30)		Female (N = 30)	
		f	%	f	%
Have you ever attended the meeting?	a. Yes	24	80	19	63.33
	b. No	6	20	11	36.67

N, Number of respondents; *f*, frequency.

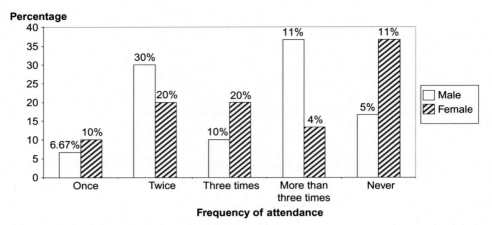

FIG. 7.1   Frequency of people in attendance at meetings.

are not allowed to speak before asked by the leader of the meeting. The *ninik mamak* claim that the practices of *adat* or *Minang* culture have shifted as a result of globalization and development in many sectors. A few members of the younger generations from Kampung Baru go to other cities to get a better education.

Although women are not constrained by customs or norms, there is no guarantee that women can express their views freely. Women are embarrassed to express their opinion because they are not used to being involved in a public meeting. Women let men decide what to do and women agree with the decision made by men. What is important for them is that water facilities are available to them. They do not think about technical problems, such as the position of the tap stands, facilities maintenance, and so on. Only a few of them participate actively, voicing their views and giving their opinions on the topic discussed. Usually, the women who participate actively have a higher level of education and better economic conditions than others.

7.3.4 Women's Participation in the Project Construction

During the project construction, men and women took part as unpaid labor to bring the material from the road to the site; to collect sand, rocks, and gravel; to dig the pipeline trenches; and other unskilled labor needed for the construction of water facilities. However, there was no clear work division between men and women. Some women were even involved in digging or lifting the pipe to a higher place. Though their presence is considered as complementary by men, women were enthusiastic to help men in the construction process.

In areas with a different gender culture, men do such work such as digging while women and children do the transporting, catering, or organizing other support activities. A different situation is found in Kampung Baru, whose people are all Minangese. The women not only do transporting or other supporting activities but also the digging, lifting the pipe, and other men's jobs. This is caused by *Minang* culture where people are used to working together, known as *Gotong royong*.

Moreover, women are used to being involved in agricultural activities, which helped them to do physical work during RWSS project construction with ease. The flexibility of women's social mobility in the context of culture has been reduced, which creates opportunities and chances for poor women to work outside their home to support their families. The same situation was also found by Joshi and Fawcett (2001a,b) in donor-funded and managed urban slum projects. *Adat* figures such as the *ninik mamak* are also involved in the RWSS project construction, particularly in determining lands or boundaries that are used for pipe trenches or tap stands. The role of *adat* figures encourages people to get involved in the activities. They feel embarrassed if they do not get involved in the activities. People's participation in project construction tends to be more a cultural matter.

Men and women are equally involved in the RWSS project construction, but a smaller number of women. In this case, gender inequality could not be observed from the payment of unskilled labors as found in many other projects that reported women were paid less than men just because of the argument that men work harder than women (Regmi and Fawcett, 2001a,b). In the study area, men and women do not get paid for their unskilled labor, only for skilled labor that usually comes from one or two individuals in the village.

During the construction work, women are used to do unskilled work, such as lifting or transporting materials. Although only a few of them were trained on technical aspects, they are capable of doing the technical work, easily understand the system, and the problems that might occur. After the project is completed and problems occur, the only thing women can do is ask for men's help or wait for the village water committee (VWC). Women should be involved in the training on technical aspects so that they can get new skills and be able to solve technical problems.

7.3.5 Women's Participation in Operation and Maintenance

In Kampung Baru, almost all of the people including men, women, and children use RWSS facilities. However, the frequency in using water facilities is quite different; men usually use the facilities for drinking but rarely use the facilities for washing their agricultural products.

Table 7.2 shows the frequency of people's involvement in the operation and maintenance of facilities. Women use the facilities more often than men do. Forty percent of women always use water facilities; 30% of them often use water facilities; 13.33% of women reported that they sometimes use water facilities; and 16.67% of them rarely use water facilities. The reason

TABLE 7.2 People's Involvement in Operation and Maintenance

	Responses	Male (N=30)		Female (N=30)	
		f	%	*f*	%
a.	Always	4	13.33	12	40
b.	Often	3	10	9	30
c.	Sometimes	14	46.67	4	13.33
d.	Rarely	9	30	5	16.67

N, Number of respondents; *f*, frequency.

why people rarely use water facilities is because they have a water source near their house. They feel more comfortable using water from a river than from other water sources. Water facilities are only used for drinking water. The frequency in using water facilities influences the task of maintenance. Women have more chance to do the maintenance task, which is mainly to keep the area around the tap clean. The maintenance task is particularly related to the damage of facilities, which is the responsibility of the VWC. People report any damage or destruction to the VWC.

Observing women's activity in fetching water, it was found that they spend more time in fetching water because they cannot bathe in the taps. Some of them are reluctant to wash their clothes at the tap stands because the location of tap stands is usually along the roadside. As a result, they have to collect more water to bring to their homes. People collected water from springs or wells two or three times a day before water facilities were built. When the facilities were built, water needs increased, but the frequency of fetching water is not significantly different from before. This is because children can bathe and collect water for their own needs.

Regmi and Fawcett (2001a,b) found a different phenomenon among women in Hille Village, East Nepal. Women complain that their water collection time increased significantly after water service was installed. Some of them have to wait until dark when they use water service to avoid being seen by men. It is important to share the burden of water collection task between men and women to save women time in collecting water. In the case of Kampung Baru, women shared the burden with their sons or daughters indirectly by the existence of the facilities. A better condition for women might be achieved if men also helped women to collect water. Consequently, women only collect water once a day. As a result, they have more time to do other activities.

The frequency of women using water facilities is related to their roles in the family. They have a responsibility to prepare food for the family, wash clothes, and bathe their children. This makes the women need more water. The different portion of water use between men and women also influences the responsibility of men and women in the maintenance of the facilities, especially for maintaining the cleanliness of the water area around the tap stands, avoiding water pollution and wastage, reducing misuses particularly from children, and diagnosing and reporting problems.

7.3.6 Women's Participation in Monitoring and Evaluation

After the RWSS project is completed and the scheme is handed over to community, the community has full responsibility to manage the water supply scheme. However, this does not mean that the project team left the scheme without any control at all. Monitoring and evaluation activities are carried out to assess the functionality and sustainability of the scheme. However, community participation in the evaluation and monitoring phase is still low. This might be caused by a low level of community educational background and lack of knowledge about the water scheme, especially regarding technical aspects. The community is not able to detect any facilities destruction. They cannot give any suggestions for a better condition of the RWSS facilities. Monitoring and evaluation conducted by donors are focused on the community satisfaction with the water supply and sanitation facilities. However, the results show that most of the people never asked about any problems during the operation of the water facilities (Table 7.3).

TABLE 7.3   Involvement in Monitoring and Evaluation Activities

Problem	Responses	Male (N=30)		Female (N=30)	
		f	%	*f*	%
Have you ever been asked by the VWC or project team concerning the constraints or problem in the operation of RWSS facilities	a. Yes	5	16.67	9	30
	b. No	18	60	14	46.67
	c. Don't know	7	23.33	7	23.33

N, Number of respondents; *f*, frequency.

People's participation in monitoring and evaluation is influenced by their participation in three previous project cycles: initiation and decision making, implementation, and operation and maintenance phases. Knowledge about the system can be derived from people's participation in previous phases. People will benefit from the system and will feel that the system belongs to them if they take part in every project phase.

7.3.7 Factors Affecting Participation in the RWSS Project

Many factors influence and form community participation in a RWSS project. Participation in the project should also consider women as well as poor and disabled people as the main water users. In this research, the factors are categorized into four main groups: formal legal system (policy), institution arrangement (project rules), sociocultural attitudes and ethic class or caste-based obligation, and religious beliefs and practices. Policy and project rules are assumed to influence community participation because institutional arrangements and policy promulgated by the stakeholders in a certain area would influence the activities conducted in every level of governance, such as RWSS activities, as one of the development activities. Sociocultural attitudes and religious beliefs are the factors pertaining to the community itself. These are embedded in the community and perceived as the local norms guiding their lives. It also influences people's behavior and how people interact with others.

7.3.7.1 Formal Legal System (Policy)

General policy and the strategy for the implementation are significantly different from the previous policies. The new policy reflects the goals in achieving project sustainability through community participation. In the implementation of the policy formulated, there is a need to have a certain strategy to ensure that the policy will be implemented in the right direction to achieve the goals of the program. The implementation of the strategy is used to clarify the general policy and to provide the real steps to implement the policy guideline in water supply and sanitation programs. However, in the strategy implementation developed by the water supply and environmental sanitation (WSES) program, there is lack of gender awareness, although it has been stated in the main policy guidelines, which explain the active role of women in the decision-making process.

It is important to understand gender equity in a development project because it is a strategic way to involve both men and women, which cannot be separated from its social relationships. In addition, women as the beneficiaries should be involved not only in the decision-making process, but also in every phase of the program, such as the implementation, operation and maintenance, and monitoring and evaluation phases. There should be clear and consistent implementation efforts to consider gender aspects. The term of a community-based development approach will be more appropriate as a gender-based development approach. A determinant factor in the implementation of a WSES program is how the agency or program executor translates the policy into action in the form of rules and regulations.

7.3.7.2 Institutional Arrangement

There are two points to be elaborated in relation to an institutional arrangement; namely, project arrangement and people involvement in organizational activities. Community involvement in the organization influences their participation in the development activities. Community involvement in organizational activities broadens their knowledge and improves their skills. They also can interact and communicate among members. At the end, this kind of activity gives indirect effect to community behaviors that are reflected in their participation in any stages of a project; the decision-making process, implementation, operation and maintenance, and monitoring and evaluation.

Community participation in the project is very much dependent on local leaders, such as the *ninik mamak,* head of *Jorong, Wali nagari,* and other older respected people in the village. Table 7.4 shows that all male respondents and 96.67% of female respondents agree that an active and an enthusiastic leader can encourage their participation in any development activities. The community considers the leaders as respected people that have higher knowledge than others. They will respect and obey a leader's rules in his governance if the leader does his duty based on the people's needs.

7.3.7.3 Sociocultural Attitudes and Ethics Class

The custom of working together improved community participation in Jorong Kampung Baru during the RWSS project construction. Some women also took part during the project construction but it was only for a certain time. Their involvement was encouraged by the needs for water. Some women in Kampung Baru are also used to working in cultivating their land. Working together is not a new habit in Kampung Baru. The role of project facilitator teams to empower and to involve the community, both men and women, is an important factor.

Social activities influence women's participation in project development activities. As a consequence, the women are used to speaking but it is only among women and not in the forum where men are involved. This kind of custom has existed and been practiced in the

TABLE 7.4   People's Perception on Local Leadership

Statement	Responses	Male (N=30)		Female (N=30)	
		f	%	f	%
An active and enthusiastic leader can encourage your participation in any development activities	a. Agree	30	100	29	96.67
	b. Disagree	0	0	1	3.33
	c. Don't know	0	0	0	0

N, Number of respondents; f, frequency.

community for a long time. This custom influences women's interaction in the project, who are not used to speaking in front of men in public.

Minang Kabau is a matrilineal society. This means that inheritance goes to women. The marriage system is exogamy, in which a man from one ethnic group (*suku*) is not allowed to marry a women from the same ethnic group. Children from the marriage will follow the mother's *suku*.

The position and roles of women in the social community tend to place women as the weak party in the community. Men see women as the ones who are responsible for doing all the household work, such as taking care of children, preparing food, collecting water, and doing other productive work in spite of their reproductive responsibilites. This situation means women do not have spare time to interact with others in the community. Women usually meet others at a certain time when they wash their clothes or take a bath in the river or at the public tap stands. They used to discuss their family issues. On the other hand, men have more time to interact with others. Men used to meet their friends after dinner. They meet to have a cup coffee or a cup of tea and then discuss their work, any issues in their village, and sometimes national issues.

7.3.7.4 *Religious Belief and Practices*

Some variables were analyzed by applying cross tabulation and a Chi-square test to see the relation between religious activities and people participation in a RWSS project. People's preference to conduct religious activity in a mosque and their involvement in religious activity were tested with people's behavior in project meetings. It indicates that people's involvement in religious activities is related to their behavior in a project meeting. The preference to conduct praying activity in the mosque indicates that women's preference for conducting praying activity in the mosque is related to their behavior in a meeting. However, most women are rarely involved in religious activities and rarely visit the mosque for praying activity.

The community comes to the mosque for praying and listening to religious talks. After the activity ends, they usually spend their spare time to interact, sharing knowledge and information with others. Project activities are one of the topics discussed. It can be concluded that the community preference for conducting prayer at the mosque or their involvement in religious activity influences their behavior in meetings, particularly related to a RWSS meeting. However, in general the study found that only a few people are involved in such religious activities. Compared to women's involvement in the activities, men have a greater frequency of engagement in religious activities. They meet each other during the Jum'at Praying at least once a week.

7.3.8 Women's Participation and Project Sustainability

Participation from all the community within the project area is needed to maintain project sustainability, in terms of technical, financial, social, environmental, and institutional areas. In addition, sustainability of the facilities by the community is also an indicator, which shows that the project not only benefits the present generation but also future generations living in the area. However, it is not an easy task to achieve the project goals. People's involvement in the project activities does not guarantee the project will achieve sustainability and benefit for the community, particularly in a RWSS project.

Women have important roles in their family, including providing safe and adequate water for the family. However, in spite of other factors that influence women's participation, some project executors or even men see women as weak and feel there is no need to involve them in the development project. In the case of the study area, most of the people who participate in the project are men. They involve a greater number of men than women and men have more power to control the activities.

Many efforts have been applied by the project executor through the community facilitator team, such as inviting women as well as men to attend a meeting and be involved in training people for project construction work. However, these efforts did not work effectively in encouraging women to participate actively. Women's presence in a meeting does not much influence the decision-making process, as they just agree with others and men have more power to control the meeting. It can be concluded that involving women in a meeting together with men cannot guarantee the presentation of the female point of view. Sociocultural factors also influence women's participation, including the educational level of the community. Moreover, other factors such as institutional arrangements, policy, and religious beliefs tend to keep women's participation at a low level. As a consequence, the sustainability of the project is affected by the lack of community awareness, particularly by women who are the main water users.

7.4 CONCLUSION AND RECOMMENDATIONS

7.4.1 Conclusion

Women's participation during the decision-making process remains low. This is due to their low educational background. Most of them do not engage in organizational activities, which could create chances for them to interact and to share information with others. Men participate more actively than women in the decision-making process. Men have more courage to speak than do women even though their education is not much higher than women. This is related to the activities men engage in. The gender equity aspect is biased during project construction. There is no clear work division between men and women. Sometimes women have to do work that is actually hard for them, such as digging and lifting the pipe. Their participation is only for the project construction phase. As a consequence, women do not have the capability to cope with problems related to RWSS facilities. Women's great will to have better RWSS facilities make them want to do all kinds of work. Sociocultural, religious beliefs, and practices are the factors that already exist in the community. These factors influence people's participation, particularly women.

Women have a high position in the *Minang* culture but they play minority roles in RWSS development activities. Women in the study area are no longer restricted by local norms because they have high mobility in productive activities. However, this does not much influence their participation in a RWSS project. Institutional arrangements at the village level in terms of organizations that exist in the community do not encourage women to participate more actively. Only a few women are involved in the village-level organization. In fact, women who engage in the organizational activities also do not make much contribution to the decision-making process.

Policies influence community participation. The policies give indirect effect to the community. Most of the respondents do not know anything about the policies in WSES. The real implementation of policies into actions is one of the factors that influence community participation. Religious beliefs and practices also do not influence women to participate. This is because most of the women are not involved in religious activities. They tend to conduct praying activities in their house and rarely pray in the mosque.

The sustainability of a RWSS project in the study area is threatened by the lack of people awareness and general knowledge about the schemes. This is due to community participation in project phases; women were not effectively involved as well as men. Lack of ownership and responsibility on the facilities leads to damage of the facilities. Some of the taps are no longer used by the community because they are located near the houses that have access to a water source. Women have to fetch water not only for drinking but also for bathing. Women cannot bathe freely in the taps because the taps are located along the roadside. Most of the people do not use public latrine facilities anymore. They feel more comfortable defecating in the streams or in the river.

7.4.2 Recommendations

Encouraging women as well as men to participate in every phase of the project is crucial to ensure project sustainability. Gaining sustainability for the project is the main goal of every project. The project is intended to give benefit to the people in the present and in the future. There is a need to think about strategy implementation in promoting the programs, which focus on gender awareness and give the people knowledge about the importance of project sustainability. It is important that the community maintain the project facilities after the project has been completed.

Involving men and women in the decision-making process has been executed by the project team. However, it was hard to achieve the goals because the community has a low level of education and a lack of self-confidence. There should be a separate meeting between men and women to encourage women to speak. Women's absence in the meeting is mainly caused by their household's activities and their roles in the family. Men allow women to be involved in a meeting. However, this does not mean that women can be encouraged to become more active participants in the project unless there is an agreement between men and women to share the work burden.

VWCs are responsible for the management of RWSS facilities. They were trained to utilize their financial, health, and technical knowledge. VWCs should create good communication with the community by conducting a meeting where all the people can express their problems. In fact, VWCs cannot dedicate their time to project management. Although there was agreement about the kinds of compensation from the water tariff, the VWCs argue that it is hard for them to manage the money because the money is only enough for operation and maintenance costs. There should be appropriate kinds of compensation that enable VWCs to give more attention to the management of RWSS facilities.

Local organizations should be organized by *adat*, religious figure, respected people, and older people in the village into a better management team. They have the capability to encourage people to join the activities to build community empowerment. This might be an effective way because the people in Kampung Baru have a great dependency on their leaders.

References

Joshi, D., Fawcett, B., 2001a. Water project and women's empowerment. In: Paper Presented at the 27thWEDC Conference, People and Systems of Water, Sanitation and Health, Lusaka.

Joshi, D., Fawcett, B., 2001b. Women and Water Engineering—A Case Study in the drinking Water Sector in Uttar Pradesh, India (Forthcoming). University of Southampton, UK.

Osmet, 1991. Village Reorganization and Irrigation Management Systems in Kecamatan Tilatang Kamang, Kabupaten Agam, West Sumatra, Indonesia. Master Thesis, Graduate School Ateneo De Manila University.

Prokopy, L.S., 2004. Women's participation in rural water supply projects in India: is it moving beyond tokenism and does it matter? Water Policy 6, 103–116.

Regmi, S.C., Fawcett, B., 2001a. Men's roles, gender relations and sustainability in water supplies: some lessons from Nepal. In: Sweetman, C. (Ed.), Men's Involvement in Gender and Development Policy and Practice. Oxfam Working Papers, Oxfam, UK.

Regmi, S.C., Fawcett, B., 2001b. Gender implications of the move from supply driven to demand driven approaches in the drinking water sector: a developing country perspective. In: Paper Presented at the First South Asia Forum on Water, Kathmandu.

Robinson, A., 2005. Indonesia National Program for Community Water Supply and Sanitation Services: Technical Guidance Report to the World Bank Indonesia and Goverment of Indonesia. World Bank, Indonesia.

Sara, J., Katz, T., 1997. Making Rural Water Supply Sustainable: Recommendations From a Global Study. UNDP-World Bank Water and Sanitation Program.

Thompson, J., Porras, I.T., Tumwine, J.K., Mujwahuzi, M.R., Katui-Katua, M., Johnstone, N., Wood, L., 2002. Drawers of Water: 30 Years of Change in Domestic Water Use & Environmental Health in East Africa. Summary. IIED, London.

WHO-UNICEF, 2004. Meeting the MDG drinking water and sanitation target: a mid-term assessment of progress. WHO-UNICEF. New York, USA

TOWARDS EFFECTIVE
MANAGEMENT OF CPRS

8

Impact of Land-Use Changes on Kuranji River Basin Functions

Fetriyuna*, Helmi†, D. Fiantis†

*University of Padjadjaran, Bandung, Indonesia †Andalas University, Padang, Indonesia

8.1 INTRODUCTION

Southeast Asia has limited land and soil resources when compared to its huge 400 million people population. Together with population pressure, land resources are deteriorating due to deforestation, uncontrolled soil erosion impact of opening land, and shifting cultivation. Land and soil along the rivers are eroded and carried away by floods or high tide. Due to deforestation, forest areas are shrinking rapidly mainly because of illegal logging, which contributes significantly to climate changes and natural disasters (floods, draught).

Land is a resource in the sense that it is capable of producing distinctive goods and services; it is also a resource because it is the spatial plane on which most human activity takes place (Field, 1997). The human-use portion of the Earth's surface has a myriad of different uses: housing, work locations, roads and other transportation corridors, farms, parks, and wilderness areas.

Land-use changes are often the result of a greatly increased population and continued economic growth and physical development, especially in a developing country like Indonesia. In this situation there is rising pressure on space and resources, which increases conflict and leads to the degradation of precious land resources. Land-use changes at the "source" or upland areas of watersheds or river basins have effects on the users of the water downstream (Susswein et al., 2001).

Land use and management techniques on watersheds/river basins can influence the quantity, reliability, and quality of water downstream. Issues of water quality and quantity are at the forefront of environmental issues in Indonesia. One of the major concerns of government, donors, and the public alike is how to manage land and water resources effectively.

A basin is the area of land that drains to a particular river or lake. Basin management is a geographically based approach of protecting and restoring water quality and quantity. This concept is also known as watershed management, particularly when applied on a smaller

105

scale (MPCA, 2002). The river basin is the logical unit of management for surface water, because water resources within the river basin are interconnected and the allocation of water in one part of the basin affects all downstream resources (UN, 2000 in MPCA, 2002).

The capacity of river basins to provide water, food, shelter, and security has been substantially reduced. Water management has moved from a sectoral approach to an integrated approach, which has now been accepted as the key for solving simultaneous challenges of securing water, food, and livelihoods, as well as protecting the environment. Water management and development is a deeply political process, which deliberately affects social as well as economic structures.

A river basin should be considered as an integrated system. Socioeconomic activities, land-use patterns, and hydromorphological processes, for example, need to be recognized as constituent parts of these systems (Global Water Partnership, 2003). Those things are believed to be interconnected with each other. The social condition will have an impact on human activities and decisions, affect land-use patterns, and finally influence the hydrological cycle in the river basin.

An important aspect of watershed or basin management is land use classification and land use planning (Cifor and FAO, 2005). It is vital that the fragile areas be identified and protected from inappropriate use, whether forestry, agriculture, or mining. However, even the "best" plan will have no impact if its implementation is not facilitated by supportive policies, a regulatory framework providing guidance, and incentive systems stimulating behavior that benefits the watershed/river basin and society at large.

The role of forests in sustaining water supplies, in protecting the soils of important watersheds, and in minimizing the effects of catastrophic floods and landslides has long been discussed and debated (Cifor and FAO, 2005). Riparian forests should be managed rigorously to protect water quality. Effective watershed and forest management consistently yields significant environmental services, including high-quality freshwater supplies. Furthermore, Kiersch (2000) argues that impacts include changes in sediment load and concentrations of nutrients, salts, metals, and agrochemicals; the influx of pathogens; and a change in the temperature regime.

Land-use practices are assumed to have important impacts on both the availability and quality of water resources. These impacts can be both positive and negative. Different land-use practices affect the hydrologic regime and water quality and at which watershed scale the impacts are of importance. For example, a change of land cover from lower to higher evapo-transpiration (ET) will lead to a decrease in annual stream flow, in tropical areas afforestation can lead to deceased dry-season flows due to increased ET, and deforestation and road construction may increase erosion (Kiersch, 2000).

Water resources and related ecosystems provide and sustain all people, and are under threat from pollution, unsustainable use, land-use changes, climate change, and many other forces (World Water Forum 3, 2003). To date, watershed and river basin management has generally achieved only partial success, largely due to the fact that biophysical factors have been emphasized at the expense of socioeconomic concerns and the fact that hydrologic boundaries are not congruent with political boundaries (Cifor and FAO, 2005).

Padang is the capital of West Sumatra Province and has an area of about 694.96 km². Padang has six river basins consisting of Batang Kandis, Batang Arau, Batang Kuranji, Batang Air Dingin, Batang Timbulan, and Batang Anak Pisang with the total area of those river basins of about 69,496 ha. Padang City is located in an alluvial plain that formed based on three major river basins (Batang Arau River, Batang Kuranji River, and Batang Air

Dingin River) with catchment area cover from Gunung Bolak, Gunung Lantik, and Gunung Bongsu. Based on its administrative boundary, Kuranji River Basin covers Padang Utara Subdistrict, Nanggalo Subdistrict, Puah Subdistrict, and Kuranji Subdistrict. The Kuranji River Basin is located at latitudes 00 48′–00 56′ and longitudes 1000 21′–1000 33′E, which exist 2.5–1600 m above sea level.

Act No. 24 of 1992 Article 7 (1) about spatial planning states the riparian zone of a river basin is a protected area and forbidden for permanent building construction, especially within a radius of 100 m from the river. But in reality, along the Kuranji River Basin and riparian zone there is a conversion and land-use changes from protected area into other uses. The problems that occur in the Kuranji River Basin are (1) increasing development and number of population, which impacts on conversion of the riparian zone into settlement area and public infrastructure; (2) cultivation practices that disobey conservation principles; and (3) disposal of industrial and domestic waste into the river, decreasing the quality of water in the river basin.

Act No. 23 of 1997 Article 1 (2) about environmental management states that environmental and natural resource management is an integrated effort to ensure the sustainability of environmental functions, which cover policies of structuring, exploitation, developing, maintaining, improving, and monitoring of the environment. Changes in technology, culture, power, and political/economic institutions all have profound influences on land use, and land-use changes along a watershed and river basin may affect hydrological functions.

This research aims to figure out the linkage between land-use changes and the capability of the river basin to maintain its functions. The main objective of the study is to address such questions as (1) what are the land utilizations in Kuranji River Basin, (2) what are the linkages between socioeconomic conditions of people upstream with land utilization in the Kuranji River Basin and forest conditions, and (3) what are the effects of land-use changes on Kuranji River Basin functions and how they affect users downstream.

This research is conducted to identify land use in the Kuranji River Basin according to a land utilization map and satellite image to study the changes, describe the linkage between socioeconomic characteristics of the people and forest condition in the upstream Kuranji River Basin, and study the land-use changes and their impacts on Kuranji River Basin functions, especially for irrigation and domestic water supply.

8.2 RESEARCH DESIGN

8.2.1 Study Area

The Kuranji River Basin was selected for this research based on the following considerations:

(1) The Kuranji River Basin is a largest river basin in Padang City, which covers 20,980.50 ha (Daus, 2005) and also has ecological, economic, and social functions.

(2) Based on research conducted by Desmiwarman (2005), the activities of the people along the river basin have impacts on the physical and chemical factors of Batang Kuranji River resulting in reduced quality and quantity of water in Kuranji River. The Kuranji River Basin provides water for irrigation, public water supply and hydroelectricity, especially for the Cement Padang factory. Gunung Nago irrigation system is the biggest irrigation

system in Padang City; it covers 4339 ha (JICA, 1983) of paddy field, which derives water from Kuranji River. Regional water supply corporation takes water from Kuranji River at Kampung Melayu. The treatment plant is located at Gunung Pangilun near the intake pump station. This plant produces $0.25\,\mathrm{m}^3/\mathrm{s}$ of water throughout the year (Fig. 8.1).

8.2.2 Research Techniques

The research is supported by both primary and secondary data. Secondary data used in the analysis are those available in *Dinas Pengelolaan Sumber Daya Air* (or Water Resources Management Agencies) (PSDA), Indonesia Center of Statistical Agencies (BPS), National Land

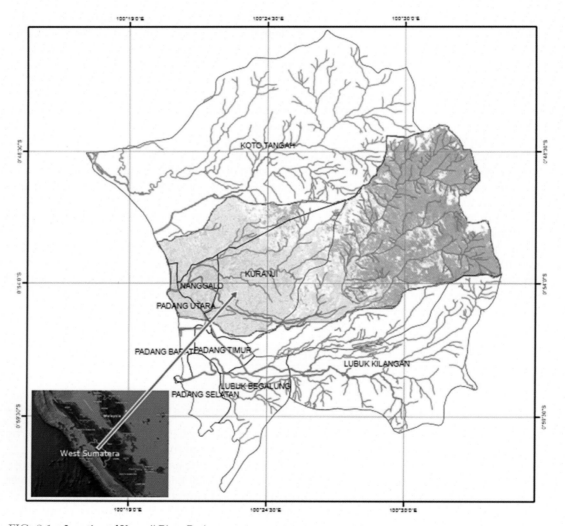

FIG. 8.1   Location of Kuranji River Basin.

Agency (BPN), household survey data that was conducted upstream in the Kuranji River Basin by Mr. Yonariza (unpublished data), and also several reports and government documents related to the research topic. Primary data gathered through in-depth interviews with key informants from the Leader of Farmer Water User Group (P3A) Limau Manis and Production Manager of Intake Pump Station (IPA) Gunung Pangilun. Secondary data is from the PSDA cover climate and the hydrological condition of the Kuranji River Basin, and also raw data of a satellite image which has been classified based on land uses. Household survey data include educational attainment, economic condition of the household, and perception about the importance of forest and forest activities. Data from BPS consist of administrative aspects for each subdistrict in the Kuranji River Basin, and the demographic condition and welfare condition based on the Village Development and Family Planning Agency in Padang City.

Data which was obtained from in-depth interviews with key informants range from the existing condition both in Municipal Water Supply Agencies and Farmer Water User Group (P3A) Limau Manis, about the continuity of supply from Kuranji River, and also conflict that has occurred in that area (Table 8.1).

Data which was obtained from in-depth interviews with key informants range from the existing condition both in Municipal Water Supply Agencies and Farmer Water User Group (P3A) Limau Manis, about the continuity of supply from Kuranji River.

8.2.3 Data Analysis

Primary data obtained through questionnaire were analyzed using a statistical method (cross tab). Demographic data and river basin condition were analyzed using the descriptive method and land cover changes were analyzed through comparative satellite images and land use maps, and finally those aspects will relate to river basin functions. Computer software such as EXCEL and SPSS (version 12) were employed for the purpose of data analysis.

Objective 1: Identify land use in the Kuranji River Basin according to a land utilization map and satellite image to study the changes.

To fulfill this objective, a satellite image and land utilization map is needed. The satellite image that was used is satellite image Landsat ETM+, year of 2002. UTM Zone47, Southern hemisphere WGS84. Land Utilization Map of Padang City, scale 1:50,000, year of 1994. The satellite image has been classified based on land-use type using ERDAS software, and the Land Utilization Map was manually digitized and grouped into land-use type using Map Info software.

Both the satellite image and the map were compared to see the changes of percentage and total area of land cover. Land use was classified as forest, mix garden, agricultural field, rice field, grass, shrub, and settlement.

Objective 2: Describe the linkage between socioeconomic characteristics of people and forest condition in the upstream Kuranji River Basin.

This objective was explored through an explanation of the demographic condition of people upstream, their educational attainment, main and secondary occupations, people's knowledge about the forest and its importance, and people's activities in the forest area related to changes of forest cover.

TABLE 8.1   Matrix Data Set

No.	Objective	Variable	Field of Data	Form	Source	Analysis
1.	Identify land use in Kuranji River Basin according to land utilization map and satellite image to study the changes	Land use changes	Land utilization of Padang City, scale 1:50,000, year of 1994	Map (manually digitized land-use map)	BPN	Compare the land cover and land utilization percentage and total area of land use
			Satellite image Lansat ETM+, year of 2002			Describe the changes through time
			UTM Zone 47, Southern hemisphere WGS 84			
2.	Describe the linkage between social economic of people and forest condition in the upstream river basin	Demographic conditions Economic conditions	Population structure Economic condition Activity in forest	- Population size - Sex distribution - Population of household - Family size - Sex ratio - Occupation - Source of income Activity of - Farming - Collect NTFP - Logging timber - Hunting animal - Others	BPS and Mr. Yonariza household survey	Statistical and descriptive analysis
		Social conditions	People knowledge about forest	The importance of forest (state and ulayat)		
3.	Study about land use changes and its impact on river basin functions	River basin function	River basin condition Climate conditions Irrigation Municipal water supply	- Stream flow - Rain fall temperature - Relative humidity evaporation - Supply for irrigation - Supply for municipal water services - Water quality	PSDA PSDA and indepth interview	Descriptive analysis

Remarks: BPN = *Badan Pertanahan Nasional* (National Land Res. Board); PSDA = *Dinas Pengelolaan Sumber Daya Air* (Water Resources Management Agencies); BPS = *Biro Pusat Statistik* (Ind. Center of Statistical Agencies.

Objective 3: Study land-use changes and their impact on river basin functions. After exploring land-use changes, this data was related with supply for irrigation and municipal water supply, and the analysis was strengthened with an explanation about climate and hydrological conditions of the Kuranji River. Explanation about the existing conditions for irrigation and municipal water supply were also obtained through in-depth interviews with the leader of the Farmer Water User Group (P3A) and Production Manager of IPA (pump intake) Gunung Pangilun. Biophysical information, including the altitudes for each subdistrict, climate conditions, and hydrological conditions, was obtained from several climatology stations, water management agencies, and also from BPS.

8.3 RESULTS

8.3.1 Land-Use Changes

Land-use pattern in the Kuranji River Basin vary between forest, agriculture area, and settlement. The upstream river basin is covered mostly by forest, whether protected forest or conservation forest (state forest), and also communal forest mainly as agroforest, which is cultivated with perennial crops for several uses. Several parts of the upstream and downstream area in this section classified as middle stream is covered by rice field area and as an urban area downstream mostly covered by settlement, industrial purposes, and commercial and institutional buildings.

Land-use changes along the Kuranji River Basin, which are derived from a comparison of satellite images (Table 8.2), show changes mainly occur between rice field and settlement.

Land-use changes are often a result of greatly increased population, economic growth, and physical development, especially in a developing country like Indonesia. In this situation there is increased pressure on space and land resources, which increase conflict and loss of capacity of resources to maintain their functions.

TABLE 8.2   Land Use Changes Basedon Land Utilization Map from Land National Board (1994) and Satellite Image From Landsat ETM+ (2002)

Land Use	1994		2002		Changes (%)[a]
	Area (ha)	(%)	Area (ha)	(%)	
Forest	12,794.00	60.98	11,119.85	53.00	−7.98
Mix garden/Agroforest	2363.67	11.27	2402.81	11.45	+0.18
Rice field	4123.99	19.66	1726.15	8.23	−11.43
Settlement	908.15	4.33	2565.15	12.23	+7.90
Shrub	77.49	0.37	1199.77	5.72	+5.35
Agriculture field	217.58	1.04	861.64	4.11	+3.07
Others (water bodies, roads, etc.)	495.64	2.36	1105.13	5.27	+2.91
Total	20,980.50	100.00	20,980.50	100.00−	

[a] *Remarks: − (means reduce); + (means increase).*

This research is an attempt to link socioeconomic conditions of the people upstream with land-use changes, especially people's consideration and priorities in utilizing their land. Based on Table 8.2, it can be concluded that

- Forest cover decreased 0.9975% per year while agroforest (mix garden) only increased 0.1225% and shrubs increased 0.6688% per year.
- Annually, rice field decreased 1.4288% whereas settlement and agriculture field increased 0.99% and 0.3838%, respectively.
- Construction of roads, increasing water body, and others increased 0.3713% annually.

Putting attention on converted forest cover into agroforest and agriculture field seems to raise pressure on the forest and may influence forest function to provide adequate supply for various users downstream. The extent to which agroforest can maintain a similar function with natural forest cover has to be investigated.

Population growth put pressure on land demand for settlement and public infrastructure. When rice field area gradually converts into settlement area, it will be in deteriorating condition in reducing total production of paddy because of reducing the area of planted and harvested rice. That situation will slowly result in food insecurity, as rice is a staple food for the majority of Indonesia's citizens.

Road construction on the one hand provides an opportunity for economic growth by increasing access to markets and public infrastructure. On the other hand, it puts pressure on land resources because it will reduce the water infiltration capacity.

The driving factors of land-use changes in the Kuranji River Basin can be classified as a combination of population pressure and socioeconomic factors, as follows:

(1) Population pressure in this area is high, with the a population increase of 37.68% in the 15 years from 1990 to 2005, with an annual growth rate of 2.41%. The growing size of the population has an impact on increasing demand of land resources for settlement and public infrastructure. In addition, increasing demand for clean and safe water, especially for domestic purposes, and increasing demand for food, which also means increasing demand for irrigation.

(2) People upstream claim to have communal forest that can be used for several purposes, especially for farming activities. They convert forest into mix garden (agroforest), which promises more income security than monoculture tree planting in an area that is claimed as communal forest.

The socioeconomic characteristics of people upstream in the Kuranji River Basin influence the condition and existence of forest.

(1) People upstream in the river basin earn their living mostly from the agricultural sector. Rice culture is cultivated near the village or on riverbanks. The product is mainly for self-consumption and the surplus is sold in the market. Besides working in the agriculture sector, people upstream in the river basin usually visit the forest for collecting nontimber forest products (NTFPs) (*asam kandis, manau,* bamboo, rattan), hunting and trapping animals, and a few of them logging timber, and also they have agroforest in communal forest. Because they depend on agriculture activities and most of their income is derived from farming activities outside the forest area, the forest condition upstream is still fairly good.

(2) People upstream already know the importance of forest for their livelihood. They consider the forest as hydrological regulation (30.0%), biodiversity conservation (20.0%), environmental protection (20.0%), a source of livelihood (7.14%), and other (12.86%). That is why people upstream maintain the existence of the forest, and by only doing farming activity in their communal forest as an agroforest.

8.3.2 River Basin Functions

8.3.2.1 Irrigation

There are two irrigation areas in the Kuranji River Basin; namely, Gunung Nago and Limau Manis. The status of the irrigation system in Gunung Nago is technical and covers a 1888 ha rice field with water demand of 4.7 L/s, while in Limau Manis the status of the irrigation area is still semitechnical and covers a 632 ha rice field with water demand of only 1.2 L/s. There is a farmers group into water user group (P3A) for each tersier channel of the irrigation system. Each P3A has the rule for water allocation based on plantation period.

Based on data from BPS in 1994, the total rice field area in four subdistricts along the Kuranji River Basin is 3696 ha, which varies from technical irrigation, semitechnical, and rain feed irrigation systems. In 2002, that number of rice field area reduced to half and became only 1726.15 ha, indicating that water demand for irrigation of rice field is reduced and can be fulfilled by the current water availability.

The amount of irrigation water available for rice field is more than enough for almost a whole year of plantation period, but there is a conflict in water allocation between paddy and fish farmers. The conflict has been mitigated with the consensus of a balanced cycle of water allocation during the plantation period.

8.3.2.2 Municipal Water Supply

Currently, only one public water supply system exists in the Kuranji River Basin with two intake pump stations (IPA), taking water from Kuranji River at Kp. Melayu. The first pump intake (IPA Gunung Pangilun) has pumping capacity of 500 L/s while pump intake IPA Jawa Gadut has pumping capacity of 20 L/s. The treatment plant is located at Gunung Pangilun near the intake pump station. This plant produces ±486 L/s of water throughout the year.

Water availability for production and finally distribution to consumers has not significantly decreased, but there is a tendency toward decreasing stream flow, yet it is still above the minimum stream flow requirement for those pump stations. The quality of water in the Kuranji River decreased significantly; however, it is still considered suitable for human consumption.

Increasing the settlement area along the Kuranji River Basin has contributed to increasing the pollution of the water in the river. While the quantity of water for domestic purposes was not significantly reduced, because the decrease of forest cover was replaced by an increase in mix garden/agroforest upstream, which has a similar function in water retention.

Water quality issues related to surface waters are now being correlated to both hydrological characteristics and terrestrial biogeochemical processes, including land-use change and other basin-wide anthropogenic issues. Land-use affect pollutant yield, both in terms of the type of pollutant generated by a particular land use and in terms of total mass of pollutant released (Black and Fisher, 2001).

8.4 CONCLUSION

Reducing forest cover is not significantly affecting river basin functions; however, it should be noted that if forest gradually decreased as much as 209.3 ha annually and only 5.56 ha is converted by agroforest per year while the rest is replaced by shrub, it will create problems when the total forest cover reaches less than 30% of total area of the river basin. This condition has already been included in a warning in Law 41, 1999, about forestry, which says that, "to maintain the environmental supporting system, forestland should be 30% of the total area of a river basin."

Road construction on one hand provided an opportunity for economic growth by increasing access to markets and public infrastructure. On the other hand, it puts pressure on land resources because it will reduce the capacity of water infiltration. There should be precise regulations on road construction to avoid the negative impact from that construction.

The results of the research show that individual's awareness and knowledge about the importance of forest play an important role to guarantee the existing and good condition of the forest. The continuity of information and a campaign about the forest functions is needed whether through electronic media and/or directly from the forestry department. Priorities must shift away from water development toward preservation and enhancement of water quality. There is a need to tie resource planning more closely to land-use planning in spite of the fact that land uses will cause an impact not only on water quantity but quality as well.

References

Black, P.E., Fisher, B.L., 2001. Conservation of Water and Related Land Resources, third ed. Lewis Publishers, Boca Raton, FL.

Cifor and FAO, 2005. Forests and Floods Drowning in Fiction or Thriving on Facts? RAP Publication 2005/03 Forest Perspectives 2.

Daus, S., 2005. Kaji ulang debit banjir rencana sungai batang Kuranji terhadap daerah korong gadang, Kota Padang (unpublished thesis). Institut Teknologi Bandung, Bandung.

Desmiwarman, 2005. Faktor Lingkungan Batang Kuranji yang mempengaruhi Kualitas Sumber Air Baku PDAM. Master Thesis, Andalas University, West Sumatra.

Field, B.C., 1997. Natural Resource Economics: An Introduction. McGraw Hill, Boston, FL.

Global Water Partnership, 2003. Poverty Reduction and IWRM. Global Water Partnership Technical Committee (TEC).

JICA, 1983. Study Report on Padang Area Flood Control Project. Main Report. Republic of Indonesia, Ministry of Public Works Directorate General of Water Resources Development.

Kiersch, B., 2000. Land Use Impacts on Water Resources: A Literature Review. Discussion Paper No. 1. Land-Water Linkages in Rural Watersheds. Electronic Workshop. Food and Agriculture Organization of the United Nations, Rome.

Minnesota Pollution Control Agency (MPCA), 2002. Basin Planning and Management an Approach to Managing Water Resources.

Susswein, P.M., van Noordwijk, M., Verbist, B., 2001. Forest Watershed Functions and Tropical Land Use Change. ASB Lecture Note 7. ICRAF SA, Bogor, Indonesia.

World Water Forum 3, 2003. Ministerial Declaration: Final Report. In Ministerial Conference on the Occasion of the Third World Water Forum (provisional edition), 22-23 March 2003. Kyoto, Japan. Available from: http:www.mlit.go.jp/tochimizushigen/mizsei/wwf3/MC-Final-Report-screen.pdf

Analysis of Incentive Factors for Sustainable Land-Use Practices: Lesson Learned From Two Case Studies in West Sumatra, Indonesia

S. Olivia Tito, Helmi, S. Husin

Andalas University, Padang, Indonesia

9.1 INTRODUCTION

Indonesia is a country rich in natural resources and biodiversity, covering 191 million ha of land, and with a population of more than 200 million people. The people of Indonesia mostly depend on the forestry and agricultural sectors for their livelihoods. Forestry offers multiple economic, social, and environmental roles, and with agriculture forms the two largest sectors of the Indonesian economy and also provides environmental services. Indonesia is highly dependent on these two sectors, and realizes the importance of proper and sustainable management of land resources.

Nevertheless, the degradation of the natural resources base that has challenged many developing countries in the world is also arising in Indonesia. Currently, the forest that once covered about 70% of the land in Indonesia is decreasing around 1.6 million ha per year as a result of deforestation (Dendi et al., 2005). Moreover, the area of degraded land in Indonesia is also increasing at a alarming rate. To illustrate, in 1984 degraded lands occupied 9.7 million ha and by 1994 this had increased to 23.2 million ha, of which 15.1 million ha is outside the forest area and 8.1 million ha is within the forest area (Mas'ud et al., 2004).

Concern over the environmental problems caused by the degradation of the natural resources base in Indonesia as well as West Sumatra has led the government to develop and encourage several rehabilitation and conservation activities. One of the forestry department's policy is "rehabilitation and conservation of land and forest resources." The policy has been implemented through various schemes. Rehabilitation and conservation is also practiced and supported by other stakeholders besides the government such as international and national nongovernmental organizations (NGOs), the private sector, and the community itself.

Moreover, centralization has changed to decentralization as the governmental systems in Indonesia and local regulations on the implementation of autonomous systems have been developed. As Government Regulation No. 9 of 2000 about a village government system (Pemerintahan Nagari), which was issued by the West Sumatra provincial government explains, the districts, subdistricts, and village governments have autonomy to manage their own resources and to develop their own regulations in managing natural resources (Boer et al., 2004). This wider autonomy for the village leader (Wali Nagari) could be seen as an important driving force for sustainable management of natural resources.

Nevertheless, diverse progress has been made in the implementation of various management initiatives in terms of sustainable land use in the community. In some cases, the progress of activities is sound as the government succeeded in applying conservation programs in the community or the community initiated conservation activities individually. In other cases, the progress of the program is slow. Overall, the variations indicates that there are factors that influence the successful implementation and development of rehabilitation and conservation initiatives. One of the major factors is the existence of incentive structures to carry out the programs.

It is important to note that natural resources use by individuals or groups is only one part of the livelihood strategy of people. The rehabilitation of natural resources degradation requires new livelihood options that change people's strategy in managing their land (Ashby, 2003). Hence, rehabilitation and conservation programs must be viewed from the perspective of farmers' overall economic goals and household strategies (Sanders et al., 1999). If farmers are rational, their land-use decisions will depend on the returns that they can get from the practice. The returns are not limited to monetary costs and benefits, but to all costs and benefits farmers derive from a given land-use practice (Pagiola, 1999).

Moreover, beside profitability a number of other factors would also influence the community's decision in practicing conservation, such as land tenure, availability of credit, technical assistance, and access to infrastructure, such as market and transportation (Liu and Li, 1999; Kamar et al., 1999; Pagiola, 1999; Bwalya, 2003; Jaireth and Dermot, 2003). Hence, some incentives are needed to encourage the community to implement the conservation and rehabilitation activity in terms of sustainable land-use practice. As Winrock International (2004) also pointed out, encouraging good stewardship of natural resources through innovative approaches, such as financial incentives, is a potential approach in promoting good land management.

Therefore, this study attempts to find incentive factors affecting the management strategy of the community in adopting a land and forest rehabilitation and conservation program in terms of sustainable land-use practices. The specific objectives are

1. To describe the existing land-use practices in the study site.
2. To identify and analyze the socioeconomic and institutional factors that form the incentive for the farmer to practice sustainable land use.
3. To draw a recommendation for the development of sustainable land-use practices at the community level.

9.2 METHODOLOGY

9.2.1 Study Area

The study area is located in two villages in West Sumatra Province. The villages are Nagari Paninggahan in Solok District and Nagari Paru in Sawahlunto Sijunjung District. These two *nagaris* were chosen due to the opportunity of have different stages and schemes of community conservation practices as part of sustainable land-use practices, which are valuable for comparative analysis. The Nagari Paninggahan has been selected because of having rehabilitation and reforestation programs from an initial stage as well as being supported by governmental and international organizations. Furthermore, Nagari Paru has a conservation initiative that is locally initiated without support from outsiders (such as government and/or international NGOs).

Nagari Paninggahan is one of the catchments areas of Singkarak Lake. The lake supports hydroelectrical power (PLTA), which provides electricity for West Sumatra and Riau provinces. The lake also provides irrigation water for the rice paddy area in many districts in West Sumatra (The districts of Tanah Datar, Padang Panjang, and Sawahlunto Sijunjung) and the land cover of Paninggahan contributes water to Singkarak Lake. However, in Paninggahan there are more unproductive lands such as grassland and critical land (about 16% of land cover) particularly in steep areas. These have an impact on the water level of Singkarak Lake that can drop by up to 2 m in dry periods (Agus et al., 2004). Thus, rehabilitation and reforestation of the critical lands for maintaining and improving environmental services are required. Furthermore, in another village Nagari Paru, communities construct the conservation as sustainable land use that not form initial stage like in Nagari Paninggahan. The communities in Paru only protect certain forest area that has been there for some time. The forest protection in Paru, usually called Rimbo Larangan, is constructed in certain catchment areas with the main goal to conserve the availability of water for the village community from that area. The Rimbo Larangan prescribes the local-based interventions that may promote effective land and forest conservation in West Sumatra. Moreover, the leader of Nagari Paru was nominated by regional government to receive the National Environmental Award (Kalpataru) of 2006 and 2007. He was nominated because of his successful support for the local initiative of natural resources conservation.

9.2.2 Research Designs

The study seeks to understand the incentive factors for the community in rehabilitation and conservation activity of land and forest resources. Therefore, it follows an exploratory and descriptive study. This is because the study describes the facts, situation, and cases dealing with land and forest rehabilitation and conservation to support sustainable land-use practices. Moreover, there is no hypothesis to be tested in this study. The major assumption about sustainable land-use practices through development of rehabilitation and conservation by the community is influenced and stimulated by incentive factors (socioeconomic and institutional factors).

Primary data, secondary data, and a process of documentation approach were used to track changes and cross-checking through and drawing on different sources of information and different approaches to recording and analyzing. The techniques in data collection include participant observation, in-depth interviews, and review of secondary data. Participant observation was applied to collect firsthand information on the events in the study site and was supposed to observe the actual condition and behavior of the local people. Thus, the triangulation technique was utilized to ensure accuracy of the information gathered, in which the insight required from the key informants through in-depth interviews were connected with the output of the observation. In-depth interviews were done repeatedly until the required information had been obtained.

The information gathered in this study comes from those who were knowledgeable in rehabilitation and conservation initiatives in the study sites, and who had been involved in those initiatives. Moreover, a reconnaissance survey was conducted in the first phase of the field study to get preinformation of the study site (profiling the village). This visit was also helpful for building relationships with the local people as well as for the planning of the field study with land users and other informants in the next phase of the field study (in-depth interviews). This study site profile was analyzed using descriptive qualitative data (ie, administrative information of the *nagari*).

9.3 RESULT AND DISCUSSION

9.3.1 Community Management Strategy in Support of Sustainable Land-Use Practices Rimbo Larangan in Nagari Paru

The Rimbo Larangan that existed in Nagari Paru is a form of local knowledge in land and forest management that is sustainable. It is an area of forbidden forest that is created and protected by the community, and exploitation of forest resources such as logging and hunting is not allowed in this area. The members of the community who violate this rule will get an *adat* (ethnic) sanction such as having to pay an in-kind fine of money or livestock. And the practice of that rule and sanction has never been implemented. There were three areas created as Rimbo Larangan. Those were commonly located in the hilly region and were a source of water for the community. There were many rivers in this area that flowed to the settlement and agriculture area of the *nagari*. The rivers are utilized by the community for agriculture and to fulfill the daily needs of water. Those three areas are Rimbo Mudik Mangan that covers ±3000 ha, Rimbo Sungai Durian that covers ±1000 ha, and Rimbo Sungai Sirah that covers ±500 ha.

The system of Rimbo Larangan was created based on initiatives to protect the source of rivers located in the hills of Nagari Paru. Droughts often arise in Nagari Paru, particularly in the dry season, because of the reduced availability of water in the rivers mainly due to increased logging in the forests. The forest should be preserved from commercial logging to maintain water availability. The idea of forest preservation had been initiated by Wali Nagari (*nagari* leader) since 1998 and some progress toward devolution of decision making over natural resources through a decentralization policy has supported Wali Nagari's idea. In the beginning, this idea faced many objections from the community, particularly from the loggers. Nevertheless, the Wali Nagari keep trying to spread awareness about his idea and build community understanding on the importance of maintaining the forest. Community awareness was built gradually by discussing the idea of forest conservation at a small store called *lapau* and every community get-together.

The idea to maintain the forest area got support from *adat* figures called *Ninik Mamak* and the community. Thus, through a public meeting called *musyawarah nagari* in 2001, an agreement among all *nagari* components (ie, community, *adat* figures, *nagari* government) was reached to protect and conserve some forest area in Paru as forbidden forest, which resulted in what was called *Rimbo Larangan*. Moreover, that agreement also includes a regulation of sanction, which for those who violate it pay a fine to the *nagari* and their *datuak* (leader of a subclan) has the responsibility as well to guarantee the payment of the fine.

To support and make the system of Rimbo Larangan more powerful there are several institutional arrangements established. For instance, the government of Nagari Paru has recognized the Rimbo Larangan system in a local regulation called *PERNA*. It was established in PERNA No. 1 of 2002 about the conservation of *nagari* forest (Rimbo Larangan) and its resources. PERNA has become a legal base for the existence of Rimbo Larangan as forest protection, followed by the creation of a written agreement by all Penghulu Sukus (ethnic leaders) in the *nagari* concerning the protection and conservation of Rimbo Larangan in Paru.

The existence of Rimbo Larangan is also supported by the construction of a local group, which has the role and responsibility to monitor the implementation of Rimbo Larangan. It consists of people who care and are aware of the forest condition. That group is called *Kelompok Petani Peduli Hutan* and has legal standing through SK Wali Nagari No. 188.47/05/Kpts-WN-2003. Furthermore, regarding the location of Rimbo Larangan bordering with the area of neighbor villages (Nagari Aie Angek and Solok Ambah), the government of Nagari Paru also discussed that system with the government from those *nagaris* to avoid any conflict. As a result, the government of neighboring *nagaris* had written a letter stating they were the representative of their community and supported the establishment and existence of Rimbo Larangan in Nagari Paru.

9.3.2 Rehabilitation and Reforestation Programs in Paninggahan

The government of Paninggahan is aware of the importance of maintaining and increasing forest cover in the *nagari*. There were a number of initiatives (programs) developed in Paninggahan as reforestation and rehabilitation activity. Moreover, the external stakeholders (national and international) are also aware of the situation and collaborate with the local government to build some rehabilitation and reforestation projects, particularly the rehabilitation of degraded land and forest area.

There has been a National Movement of Land and Forest Rehabilitation (GNRHL) program in Paninggahan since 2004, as well as other funding schemes of land and forest rehabilitation such as from *Dana Alokasi Khusus Dana Reboisasi* (DAK-DR) (from *Anggaran Pendapatan dan Belanja Daerah* (APBD); that is, provincial income). Eventually, there was a Friendship Forest with a funding scheme from the Japan International Forestry Promotion and Research Organization (JIFPRO) in 2004–05. In spite of this, the local government still attempted to find and develop other proposals for the next year to handle another degraded land in Paninggahan. For instance, in 2006 the local government through the *nagari* institution for environment *Badan Pengelola Lingkungan Hidup* (BPLH) tried to develop a proposal for another friendship forest or to apply the Clean Development Mechanism concept in Paninggahan as well an environmental services program.

The activities of the GNRHL movement in Paninggahan consist of community forestry and reforestation. It is located mostly in Jorong Subarang and Jorong Koto Baru Tambak, and totally covers almost 225 ha. Moreover, land and forest rehabilitation covered 75 ha from the APBD budget through DAK-DR. The community forestry is a GNRHL activity that attempted to rehabilitate the critical land for recovering the function and increasing the productivity of land through a variety of plantings (timber and nontimber). It provides an opportunity to increase the community welfare and recover the environment as well. Likewise, reforestation is an activity that attempts to recover the function of a degraded forest area. The GNRHL in Paninggahan is a part of land and forest rehabilitation in the catchments area of Singkarak Lake that is a priority lake (DAS Agam Kuantan); thus, the government provides incentives for all the rehabilitation activities. Moreover, the farmer also gets a daily wage for each activity in GNRHL called *Harian Ongkos Kerja*. The payments are made directly to the account of a working group through the system of *Surat Perjanjian Kerja Sama*; that is, a cooperation agreement letter.

There is 50 ha of friendship forest in Paninggahan that resulted from rehabilitation and reforestation activity. The funding for the activity was facilitated by JIFPRO with a contribution from Wakanyaku Medical Institute Ltd, Japan. That activity was located in Bukit Batu Agung, Jorong Subarang, which was carried out by 43 farmers. There are a variety of trees planted in that activity such as fruit and wood trees (eg, *tectona grandis, persea americana, aleuritis moluccana, eugenia aromatica, shorea sp*, etc.). This program provides the farmer with free seeds, funds for planting, fertilizer, and a two-year maintenance fund.

The establishment of friendship forest is a result of the people's awareness of the need solve the degradation and environmental problems that were on the rise in Paninggahan, which is an important catchment area of Singkarak Lake. The *nagari* government and community are actively trying to find and create rehabilitation activities such as this friendship forest. The process to establish the friendship forest begins with an offer of a proposal to JIFPRO by the *nagari* institution for environmental BPLH. JIFPRO observed the area and signed the agreement with Nagari Paninggahan government for the friendship forest after conducting some capacity building.

Based on the observation in the field area of friendship forest and discussion with the JIFPRO officer, the progress of the activity is running well. A 75% seed germination rate in a stony land for the second year indicates that the project is yielding the promised results. The success of the current project satisfied the JIFPRO and raised their interest in implementing another program of rehabilitation of land and forest in Paninggahan.

9.3.3 Similarities and Differences of Incentive Factors

One aspect of the two rehabilitation and conservation cases in this study is the same. Both cases have some socioeconomic, policy, and institutional factors that form incentives for the community to initiate and develop rehabilitation and conservation practices. The incentives motivate and stimulate the community to take appropriate management of their land and forest resources to solve environmental and natural resources problems arising in both *nagari*s. Those incentive factors relate to each other to stimulate rehabilitation and conservation initiatives.

While a broad similarity exists in the two cases, this study also identifies some differences (Table 9.1). The major difference of the incentive factors for the rehabilitation and conservation initiatives in both cases is from the actors involved. In the Paru case the conservation initiative involves the local community only. There is no assistance or support from other stakeholders in establishing as well as maintaining the existence of Rimbo Larangan. Nevertheless, in Paninggahan several stakeholders from regional, national, and international levels contribute to initiate and develop rehabilitation and reforestation programs. Government as well as international NGOs are concerned about rehabilitation in Paninggahan, and they provide several incentives for the local community to implement the program.

TABLE 9.1   Similarities and Differences of Incentive Factors in Paru and Paninggahan

Incentive Factors	Similarities	Differences
Knowledge and awareness	Community aware on environmental problems that take place	In Paru, awareness develops by insiders (local community), and in Paninggahan awareness is encouraged by outside stakeholderd (includes government, national and international NGOs)
Benefit	Beneficial for the community	In Paru, the benefit is the improvement of productivity in wet rice field, and in Paninggahan the benefit includes the opportunity of income through cultivation of intercrop
Material, financial, and technical incentives	There are no similarities from these kinds of incentives	There is no material, financial, or even technical incentive for the community in Paru. Nevertheless, in Paninggahan cost for the practice is subsidized, which consists of planting and maintenance costs and providing planting material and technical guidance.
Easy access to service and infrastructure	There are no similarities from these kinds of incentives	Available and easy access to local market in Paru for agriculture products supports the community to keep and improve their agriculture activity rather than exploit the forest such as by logging. In Paninggahan, although there is easy access to services and infrastructure, there is no direct contribution to or influence on the people's decision to be involved in a rehabilitation program.
Land tenure system	Having a strict land tenure system is the customary system	In Paru, all locations of Rimbo Larangan belong to communal land (clan land). Nevertheless, in Paninggahan land included in rehabilitation and reforestation activity not only belongs to the clan (*ulayat suku*), but there it also belongs to the *nigari* (*ulayat nagari*) and subclan (*ulayat kaum*).
Policies and institutional factors	Consists of enabling policies, local regulation toward rehabilitation and conservation activity, and local organization formation	• Institutional arrangements are initiated and developed by local people, but in Paninggahan this is assisted by outside stakeholders from the regional, national, and international levels (such government, NGOs, academicians). • Local organizations in Paru work more in operational ways (physical monitoring), and in Paninggahan more in governance ways (in decision making).

TABLE 9.2   The Consequences of Doing or Not Doing Rehabilitation and Conservation Initiatives in Paninggahan and Paru

Paru Case		Paninggahan Case	
Doing	**Not Doing**	**Doing**	**Not Doing**
Improving water supply of river to fulfill *sawah* and daily needs	Reducing water of river to fulfill *sawah* and daily needs, particularly in dry periods	Increasing area of productive lands	Increasing area of unproductive land (critical land)
Improving productivity of *sawah*	Reducing productivity of *sawah*	Increasing area with forest	Reducing forest area
Reducing logging	Increasing logging	Maintaining and increasing water quality and quantity of river and lake, particularly in dry periods	Decreasing water level of Singkarak Lake; reducing water irrigation, particularly in dry periods
Biodiversity conservation	Loss of biodiversity	Income generation opportunity through intercropping	Loss of income generation opportunity
Risk reduction of natural disasters (such as landslides)	Increasing risk of natural disasters (such as landslides)	Risk reduction of environmental problems that control landslides, reduce loss of fertile soil through erosion	Increasing risk of environmental problems such landslides and erosion
Provide beautiful landscape (possibly develop as tourism objects)	Loss of opportunity incomes from tourism development	Receive funding for the cost of practices	Loss of funding opportunity

Thus, based on the explanation above, the successful factors for sustainable land-use practices at the local level includes the existence of local rules, a functioning local organization, and an ability to enforce the rules (effective rule enforcement in the *nagari*) also recognizing management initiatives as well as acknowledgment from the government.

Furthermore, there is a question of why the community should initiate and or practice rehabilitation and conservation activities. To answer that question it is necessary to explain what will be the outcomes or benefits if they do and if they not do (Table 9.2). Opportunity costs for doing or not doing rehabilitation and conservation activities should be shown.

9.4 GENERAL DISCUSSION

Learned from the case study in Paru and Paninggahan, the incentives factors have motivate the community to establish and develop the rehabilitation and conservation activities. It has stimulated the development activity and motivational changes of the community to encourage and maintain proper land management. Therefore, this presents lesson learned from the case study of the important factors that influence, determine and motivate the development of sustainable land use practices.

9.4.1 The Importance of Knowledge and Awareness of the Local Community About Natural Resources Problems and Benefits

If the local community is aware of the natural resources degradation and environmental problems (causes and effects) that have threatened their livelihoods, this knowledge may motivate them to change their practices into proper management activity. It would support community involvement in the development of strategies and actions for proper land management and related resources (Hannam, 1998; Holden and Shiferaw, 1999). Building the local community's knowledge and awareness on the problems and how to manage them should be performed in several ways to familiarize the community and motivate them to implement a proper management strategy (Palmer et al., 1999). Practices that are profitable and simple to implement are more likely to be implemented (Pagiola, 1999; Liu and Li, 1999; Nabben, 1999; Nowak and Korsching, 1983; cited in Klapproth and Johnson, 2001).

9.4.2 Material, Financial, and Technical Support for Sustainable Land-Use Practices

Availability of the material, financial, and technical incentives have increased the community's willingness to become involved in rehabilitation and reforestation activity. The subsidized cost for all programs would generate more financial benefits to the local communities and motivate them to be involved in rehabilitation and reforestation programs. The local communities have limited capital to practice management strategies; the financial support for those activities will make it easier for them to begin the practice (see, for example, Scherr and Current, 1999; Hannam, 1998; Kamar et al., 1999; Scherr and Current, 1999). Moreover, easy access of a support mechanism is also important for sustainable land-use practices. Easy access to services and infrastructure such as the availability of a financial institution, markets, and road access would also influence the motivation of the local community. This is the development that creates an enabling environment for the local community and encourages them to develop and adopt management practices (Kamar et al., 1999).

9.4.3 Clear Land Tenure Arrangement

A clear land tenure arrangement will provide security for the land and reduce the constraint for development of some policy instrument of the land. Most customary rules found are able to provide s clear and strict land tenure arrangement. Thus, it may reduce the complexity in establishing and developing of some proper management practices of natural resources and environment, as well as to maintain the existence of the practices (see also Palmer et al., 1999; Phien and Tu siem, 1997, cited in Sanders and Cahill, 1999; Pagiola, 1999; De Foresta et al., 2000).

9.4.4 Functioning of Local Institutions and Resource Mobilization Mechanisms for Sustainable Land-Use Practices

In many cases, governments do not have the capacity to manage the natural resources in their jurisdiction exclusively and effectively. Thus, transferring this capacity to the local community by strengthening local institutions is a good idea. Given the opportunity for the local community to manage their natural resources may well be able to identify underlying problems and work out possible solutions. Thus, functioning local institutions may stimulate the development activity and motivate changes of the local community over proper natural resources management.

The importance of functioning local institutions and resource mobilization to the local community over their natural resources management have been found in several cases. The case of Rimbo Larangan in Paru found the creation of an institutional arrangement and enabling regulation over proper forest management resulted through a functioning local institution. In Paninggahan, with resource mobilization over environmental and natural resources management to the local community, the local institutions function for developing proper management (see also Palmer et al., 1999; Nabben, 1999).

9.4.5 Government Policy That Provides Room and Encouragement for Local Initiatives for Sustainable Land-Use Practices

Better government policies and enabling legislation are certain prerequisites that are important to use incentives. It is necessary because it provides room and may encourage the farmer to adopt more sustainable practices. The national government policy of decentralization has provided district/municipal government with responsibilities for their natural resources management. Act No. 22 of 1999 and Government Regulation No. 9 of 2000 about the village government system (Pemerintahan Nagari) has provided room for *nagari*s to manage their own natural resources and to develop their own regulations for controlling them. It has provided an opportunity for local initiatives to manage their own natural resources through their own decisions, based on their own aspirations, which is officially permitted.

Furthermore, the national government has policy instruments to involve the local community in managing the forest. There was a national policy that recognized community-based management of natural resources. In line with the land and forest rehabilitation that has become a priority policy of forestry development by the national government, the local initiatives for forest management have an opportunity to develop. Thus, the policies and legal recognition of local rights to manage their natural resources could be seen as an important driving force for sustainable management of natural resources.

9.5 CONCLUSIONS AND RECOMMENDATIONS

9.5.1 Conclusions

1. The community in Paru and Paninggahan has able to demonstrate capacity and initiatives in developing sustainable land-use practices. It was illustrated in several rehabilitation and conservation initiatives as proper management of land and forest resources. The conservation initiative in Paru through Rimbo Larangan is established and maintained only by the local community. The activity is locally initiated without support from outside stakeholders (government and/or international NGOs). The rehabilitation and reforestation initiatives in Paninggahan are developed and financially supported by governments and outsiders (international funding organizations).

2. This study highlighted the importance of incentives for the local communities to initiate or develop activities to solve degradation problems. It affects the behavior, interest, and adoption of the community in the initiatives of rehabilitation and conservation. The incentives are mentioned below.

3. The knowledge and awareness of the community of the causes and impacts of the land and forest degradation problem have motivated them to become involved in the development of strategies and actions for proper land management and related activities.

4. Therefore, the socialization and education of environmental problems and the impacts may help familiarize the community with the need for rehabilitation and conservation practices.

5. Benefits that can be obtained by the community in rehabilitation and conservation practices are one of the reasons for them to be involved. The practices that are profitable and simple to implement are more likely to be implemented.

6. Subsidization of the rehabilitation programs have modified the benefits that can be obtained by the community. Consequently, the people have a stronger view of their involvement in rehabilitation and reforestation programs, and the adoption of that practice is high. Subsidization and material, financial, and technical incentives along with access to infrastructure and services are necessary in the early stage of rehabilitation and conservation.

7. Most of the customary rules found are able to provide a clear and strict land tenure arrangement. Secure land tenure has great effect on the adoption of proper land management practices.

8. Environmental awareness has supported institutional development for rehabilitation and conservation initiatives, which includes functioning local institutions.

9. Local initiatives included for the development of rules and a legal framework (regulation) are necessary to implement the initiatives, as well as the creation of institutions that facilitate the implementation. The institutional arrangement has stimulated the development of activities and motivated the community to maintain rehabilitation and conservation practices.

10. Legal recognition of the local right to manage their own natural resources (decentralization), which provide local government wider autonomy, could be seen as an important driving force for sustainable management of natural resources. This authority is describe as a government policy that may provide room and encouragement for local initiatives of sustainable land-use practice.

9.5.2 Recommendations

1. Enhance and raise public awareness of various environmental issues, the importance of proper management, and the capability of local participation in addressing these issues. Provide clear information about the benefits of participation and the consequences of no participation in rehabilitation and conservation activities is also necessary. Building awareness of the local communities through training and print and electronic media may raise the community's understanding of conservation and rehabilitation.

2. The financial support needed to carry out the rehabilitation and conservation program is also important because it can enlarge the benefits that can be obtained by the local communities involved. Moreover, clear information of the benefits and or financial support available through training will improve the knowledge of people about the advantages to be gained from rehabilitation and conservation initiatives.

3. Functioning of the local institutions to initiate proper land management practices is important. Thus, it is necessary to develop and maintain high local participation through implementation of the community-based natural resources management approach in every rehabilitation and conservation program. Besides the contributions from donor organizations and government in management activities, empowerment of the local people or local institutions is also necessary. Developing policy support, including a legal framework to facilitate and control the management strategy, institutional capacity building, and strengthening the institutions and or local organization are essential in providing a positive environment for rehabilitation and conservation programs.

4. Institutional capacity building or strengthening in each area is different as regards local needs and aspirations. For instance, in some villages, due to their low capacity and lack of experience, the people would not be able to write a new law of management. Thus, concern about local aspirations and needs in institutional capacity building and strengthening is necessary.

References

Agus, F., Farida, F., van Nordjwik, M. (Eds.), 2004. Hydrological Impacts of Forest, Agroforestry and Upland Cropping As a Basis for Rewarding Environmental Services Providers in Indonesia. Proceedings of a Workshop in Padang/Singkarak, West Sumatra, Indonesia, 25–28 February 2004. ICRAF-SEA, Bogor, Indonesia.

Ashby, J., 2003. Introduction. In: Pound, B., Snapp, S., McDougall, C., Braun, A. (Eds.), Managing Natural Resources for Sustainable Livelihoods: Uniting Science and Participation. London, UK, Earthscan.

Boer, R., Bulrk, A., Djisbar, A., 2004. Challenges and opportunities to implement RUPES program at Singkarak catchment. In: Agus, F., Farida, F., van Noordwijk, M. (Eds.), Hydrological Impacts of Forest, Agroforestry and Upland Cropping As a Basis for Rewarding Environmental Services Providers in Indonesia. Proceedings of Workshop in Padang/Singkarak, West Sumatra, Indonesia, 25–28 February 2004. ICRAF-SEA, Bogor, Indonesia.

Bwalya, S.M., 2003. Understanding community-based wildlife governance in Southern Africa: a case study of Zambia. AJEAM-RAGEE 7, 41–60.

De Foresta, H., Kusworo, A., Michon, G., Djatmiko, W.A., 2000. Ketika Kebun Berupa Hutan: Agroforest Khas Indonesia, Sebuah Sumbngan Masyarakat. ICRAF, Bogor, Indonesia.

Dendi, A., Shivakoti, G.P., Dale, R., Ranamukhaarachchi, S.L., 2005. Evolution of the Minagkabau's shifting cultivation in the West Sumatra highlands of Indonesia and its strategic implications for dynamic farming systems. Land Degrad. Dev. 16, 13–26.

Hannam, I.D., 1998. Soil conservation policies in Australia: successess and failure and requirements for ecologically sustainable policy. In: Napier, T.L., Camboni, S.M. (Eds.), Soil and Water Conservation Society. 7515 Northeast Ankeny Road, Ankeny, Iowa, USA, pp. 618–638.

Holden, S.T., Shiferaw, B., 1999. Incentives for sustainable land management in peasant agriculture in the Ethiopian Highlands. In: Sanders, D.W., Huszar, P.C., Sombatpanit, S., Enters, T. (Eds.), Incentives in Soil Conservation: From Theory to Practice. Oxford & IBH Publishing Co. PVT. Ltd, New Delhi, pp. 275–294.

Jaireth, H., Dermot, S. (Eds.), 2003. Innovative Governance: Indigenous Peoples, Local Communities and Protected Areas. Ane Books, New Delhi.

Kamar, M.J., Mburu, J.K., Thomas, D.B., 1999. The role of incentives in soil and water conservation in Kenya. In: Sanders, D.W., Huszar, P.C., Sombatpanit, S., Enters, T. (Eds.), Incentives in Soil Conservation: From Theory to Practice. Oxford & IBH Publishing Co. PVT. Ltd, New Delhi, pp. 231–246.

Klapproth, J.C., Johnson, J.E., 2001. Understanding the Science Behind Riparian Forest Buffers: Factors Influencing Adoption. Virginia Cooperative Extension Virginia Tech, Blacksburg, VA.

Liu, A.-P., Li, Y., 1999. A case study of using soil conservation incentives in the Chinese loess plateau. In: Sanders, D.W., Huszar, P.C., Sombatpanit, S., Enters, T. (Eds.), Incentives in Soil Conservation: From Theory to Practice. Oxford & IBH Publishing Co. PVT. Ltd, New Delhi, pp. 215–230.

Mas'ud, A.F., Nugroho, C., Pramono, I.B., 2004. Criteria and indicators of watershed management used for the national movement for land and forest rehabilitation (GNRHL) in Indonesia. In: Agus, F., Farida, F., van Noordwijk, M. (Eds.), Hydrological Impacts of Forest, Agroforestry and Upland Cropping As a Basis for Rewarding Environmental Services Providers in Indonesia. Proceedings of Work shop in Padang/Singkarak, West Sumatra, Indonesia, 25–28 February 2004. ICRAF-SEA, Bogor, Indonesia.

Nabben, T., 1999. Funding to Community Landcare Groups in Western Australia. In: Sanders, D.W., Huszar, P.C., Sombatpanit, S., Enters, T. (Eds.), Incentives in Soil Conservation: From Theory to Practice. Oxford & IBH Publishing Co. PVT. Ltd, New Delhi.

Nowak, P.J., Korsching, P.F., 1983. Social and institutional factors affecting the adoption and maintenance of agricultural BMPs. In: Schaller, F., Bailey, G. (Eds.), Agricultural Management and Water Quality. Iowa State University Press, IA, Ames, USA, pp. 349–373.

Pagiola, S., 1999. Economic analysis of incentives for soil conservation. In: Sanders, D.W., Huszar, P.C., Sombatpanit, S., Enters, T. (Eds.), Incentives in Soil Conservation: From Theory to Practice. Oxford & IBH Publishing Co. PVT. Ltd, New Delhi, pp. 41–56.

Palmer, J.J., Guliban, E., Tacio, H., 1999. Use and success of incentives for promoting sloping agricultural land technology in the Phillippines. In: Sanders, D.W., Huszar, P.C., Sombatpanit, S., Enters, T. (Eds.), Incentives in Soil Conservation: From Theory to Practice. Oxford & IBH Publishing Co. PVT. Ltd, New Delhi, pp. 309–324.

Pound, B., Snapp, S., McDougall, C., Braun, A. (Eds.), 2003. Managing Natural Resources for Sustainable Livelihoods: Uniting Science and Participation. Earthscan, London, UK.

Sanders, D., Cahill, D., 1999. Where incentives fit in soil conservation programs. In: Sanders, D.W., Huszar, P.C., Sombatpanit, S., Enters, T. (Eds.), Introduction: Incentives in Soil Conservation: From Theory to Practice. Oxford & IBH Publishing Co. PVT. Ltd., New Delhi, pp. 11–24.

Sanders, D.W., Huszar, P.C., Sombatpanit, S., Enters, T., 1999. Introduction. In: Sanders, D.W., Huszar, P.C., Sombatpanit, S., Enters, T. (Eds.), Incentives in Soil Conservation: From Theory to Practice. Oxford & IBH Publishing Co. PVT. Ltd, New Delhi, pp. 1–7.

Scherr, S.J., Current, D., 1999. Incentives for agroforestry development: experience in central America and the Caribbean. In: Sanders, D.W., Huszar, P.C., Sombatpanit, S., Enters, T. (Eds.), Incentives in Soil Conservation: From Theory to Practice. Oxford & IBH Publishing Co. PVT. Ltd, New Delhi, pp. 345–365.

Winrock International, 2004. Financial Incentives to Commuities for Stewardship of Environmental Resources Feasibility Study. Submitted to U.S. Agency for International Development and Asia and the Near East Bureau/Office of Technical Support, Virginia, USA. Available online: http://www.winrock.org/fnrm/files/FinancialFINAlrev.pdf.

Forest Management and Illegal Logging in West Sumatra: The Case of Sangir, South Solok

Yolamalinda*, S. Karimi†, R. Febriamansyah†

*STKIP PGRI, West Sumatra, Indonesia †Andalas University, Padang, Indonesia

10.1 BACKGROUND

Over the last two decades of the 20th century, rapid deforestation caused the loss of around 15 million ha of forest in the world, largely in the tropics. There are a complex set of forces causing forests to decline. One of the most intractable problems that societies have to face is the scale of corruption in the forest industry. Illegal practices jeopardize the financial capacities of governments and societies to sustain both forests and development (Krishnaswamy, 1999).

Illegal logging can be defined as the harvesting of logs in contravention of laws and regulations. These laws and regulations were designed to prevent the overexploitation of forest resources and to promote sustainable forest management. In accordance to this definition, illegal activities may include logging in protected areas, the logging of protected species, logging outside concession boundaries, extraction of more than the allowable harvest, removal of oversized or undersized trees, and harvesting in areas where extraction is prohibited such as catchments areas, steep slopes and river banks, export and import of illegally harvested timber without paying taxes, and use in processing (AFLEG, 2003; Casson and Oidzynski, 2002).

There are essentially two kinds of illegal logging. Legitimate operators who violate the terms of their licenses carry out the first. The second involves outright timber theft, whereby people who have no legal right to cut trees do so anyway. In the current decentralization era, more types of illegal logging is carried out under the auspices of illegally obtained harvesting permits issued by a district government official (Barber, 2002; Casson and Setyarso, 2005).

Many research studies all over the world include the socioeconomic situation in rural areas and legislative problems as factors that support illegal logging. Low-income level, unemployment, and crisis in agricultural sectors are some of the important socioeconomic situations that contribute to an environment of illegal logging practices. In terms of legislative problems, contradictions between legal acts and regulations and complexity of legislation are common (WWF, 2004; Casson and Oidzynski, 2002; Mc. Carthy, 2002).

Illegal logging is a complex problem and there are multiple causes for its existence in Indonesia. Some of the key causes of illegal logging identified through extensive consultations include overcapacity of the wood processing industry; domestic and international demand for illegal timber; systemic corruption; and rent-seeking behavior, rapid decentralization, growing unrest with the status quo, and poor law enforcement.

"Illegal" logging, particularly by local communities, is not a new phenomenon in Indonesia. Tensions between the state and local interests over the control of forest resources in "outer island" Indonesia have a history that extends back to the colonial period. The policies of the New Order merely sharpened and intensified these tensions. Under the New Order regime, all Indonesian forests were declared state forests and the outer island forests were opened to large-scale timber extraction to generate much needed revenue. This centralized control of Indonesia's forestry sector followed the 1967–70 period of a relatively relaxed policy that allowed district authorities and village communities to engage in small-scale logging activities. However, they were marginalized in favor of multinational corporations with a link to the central government. The centralization of natural resources management has led to a loss of community participation (Casson and Oidzynski, 2002).

Casson and Oidzynski (2002) and Mc. Carthy (2002) considered the economic and political crisis as driving factors for illegal logging in Indonesia. After a crisis hit Indonesia in mid-1997, many local people lost their jobs in the manufacturing and industrial sectors. They began to rely on the forest sector to meet their daily needs. Despite an overall drop in international timber prices in Japan, Korea, and the China market—the price for roughly sawn timber in the main transshipment states of Sarawak and Sabah remained attractive for Indonesian sellers (US $250) in the year 2000. Shipping sawn timber internally primarily to Java, although comparatively less profitable at US $120, was also considered to be worthwhile. After the fall of Soeharto, regulation reform was highly concerned with the need to adopt community-based forest management that considers the tenure rights of indigenous people and local communities to forestland and resources (Siscawati, 1999). The changing regulatory atmosphere at the *reformasi* issued instructions allowing communities in or near forest areas to be actively involved in forest exploitation, take a role in forest management, and to be allowed to carry out low-impact extraction activities, primarily of nontimber forest products (NTFPs), but it did not work out as expected and illegal logging still happened. Communities much prefer to harvest timber rather than NTFPs (Casson and Oidzynski, 2002; WRM, 2005).

This chapter has three objectives: (1) to identify illegal logging practices, (2) to identify socioeconomic characteristics of the community, and (3) to identify major factors contributing to illegal logging. This study focuses on villagers as one of the actors in illegal logging. It is also aimed at investigating factors contributing to illegal logging.

10.2 METHODOLOGY

South Solok was selected as the research site after consulting some key informants from the forestry department. The district is known as Sangir and the lowest governance is Lubuk Gadang. There are 22 jorong in Lubuk Gadang, but due to time limitations only three *jorong*s were selected: Padang Aro, Lubuk Gadang, and Durian Taruang. These three *jorong*s were selected because the majority of their people depend on forest resources for their livelihoods.

The sources of primary data were rapid rural appraisals, household questionnaire surveys, and key informants interviews. Household surveys were conducted to gather primary data on the socioeconomic characteristics of the sampled households. The survey questionnaires were designed to collect the information on socioeconomic characteristics and perception of the community on illegal logging. The local elite, forest office personnel, and related NGOs were consulted for the cross-checking and validation of received responses. Semistructured interviews were conducted to interact with the second category of informants. Direct observations and transect walks with local people were applied to get more reliable information from the sites. The secondary data were collected from the forest office, statistics bureau, *nagari* office, website, and the attorney's office.

10.3 FINDINGS AND DISCUSSIONS

10.3.1 Forest Management

The idea of forest management is to recognize its ecological, social, and economic functions. Forests also provide economic benefits for local communities such as jobs and income. *Garis Besar Haluan Negara* of 1983 introduced the integrated approach to forest management to maintain the sustainability of whole elements of the environment. The Forestry Act of 1999 sets out principles of forest management including utility, sustainability, democratization, justice, commonness, transparency, and integration.

During the Soeharto era, all Indonesian forests were classified as state forest as a result of a government plan to increase government revenue. Through Act no. 41 of 1999 on forestry, Chapter 6 Verse 2 indicates that forest in Indonesia are divided into three categories: (1) production forest, (2) conservation forest, and (3) protected forest. However, according to *Tata Guna Hutan Kesepakatan* (forest land-use policy) (TGHK) of 1999, forests are divided into six categories: (1) preservation forest, (2) protected forest, (3) limited production forest, (4) conversion production forest, (5) production forest, and (6) other use area. This distribution on West Sumatra forest is shown in Table 10.1.

Forests, covering 38,784 ha in Sangir, are rich in many species of wood and minerals that are valuable in the domestic and international market. The forests are owned by the state, and are used as limited production forests according to TGHK. There are three major types of interaction between village people and the forest in Sangir that tend to have negative impacts on the sustainability of the forest. The first is cultivating the land inside the forest; the second is mining of gold and tin. The Dutch-based gold mining companies have already left because there is little gold left; so have the tin companies. The gold,that is left is exploited by villagers

TABLE 10.1   Forest Function According to Forest Land-Use Policy (TGHK) in West Sumatra

No.	Forest Function	Area (Ha)
1.	Preservation forest	599,694
2.	Protected forest	1,206,624
3.	Limited production forest	539,707
4.	Production forest	596,844
5.	Conversion production forest	437,733
6.	Other use area	849,128
Total		4,229,730

with traditional tools. The third interaction is harvesting forest product, particularly the illegal extraction of timber. There are two categories of people involved in the illegal logging; namely, poor people who hope to earn income to fulfill their needs and wealthy people who hire local poor people for illegal logging activities.

Wood from forests was only used for making houses and as a fuel in Sangir. Local people harvest forest products with traditional tools such as an axe. This pattern gradually changed when villagers started using chainsaws for the purpose of harvesting timber, which increased logging activities. Chainsaws became famous in Sangir since 1984. It was the same era when the price of oil dropped in 1982 to make the forestry sector the second largest contributor to foreign exchange in Indonesia after oil and gas. Wood-based industry has grown rapidly in Indonesia accompanied by the increasing demand for wood as a pulp material. Logging activity in the whole Indonesian forest increased, and according to key informants, logging activity in that era also increased in Sangir. Construction of roads from village to forest build by the HPH company also accelerated logging activities.

10.3.2 Illegal Logging Practices

The abundance of forest resources attract people to harvest forest products for financial benefits. The financial needs of people living around the forest pushed them to extract timber. Illegal loggers also influence other villagers to work as illegal loggers. The process of illegal logging is explained by reviewing the following factors.

10.3.2.1 Selection of Trees

Villagers choose trees that are of good economic value in domestic and international markets. Forests in Sangir are rich in valuable species including Meranti (*Shorea* spp.), Keruing (*Dipterocarpus* spp.), Mersawa (*Anisopteracostata*), Timbalun, and Banio (*Shorea* spp.).

10.3.2.2 Villagers

After choosing valuable and good quality trees, logging begins. According to key informants, there are no conflicts among the loggers in choosing trees. The rules of conduct between them is "who come first they get the tree." The illegal loggers then hire laborers who

TABLE 10.2   Illegal Logging Practices and the Labor Involved

No. (Rp)	Activity	Number of Illegal Loggers	Wage/Cubic
1.	Chainsaw	1	1,000,000
2.	Hauling	4	500,000
3.	Transportation	1	500,000

Source: Field survey.

TABLE 10.3   Comparison Between Illegal Logging Wage and Farming

No.	Activity	Wage/Day (Rp)
1.	Chainsaw	142,857
2.	Hauling	71,428
3.	Transportation	71,428
4.	Farming	15,000

Source: Field survey.

use chainsaws to cut the timber down and cut it into blocks. Illegal logging may involve at least six laborers to carry the blocks from the forest to the road through the river and load it onto trucks (Table 10.2).

(Table 10.3) is a comparison of the wages community members get from forest and farming. The table shows that working in the forest as an illegal logger gives significantly higher revenues compared to farming activities. Higher earnings make people prefer logging to farming.

10.3.2.3 Tauke Kayu (Middleman)

Tauke kayu are the middlemen who buy blocks from villagers. They collect the wood and transport it to cities. Some of the middlemen also run illegal sawmills to turn the logs into blocks, and snatch higher profit margins. One illegal sawmill can employ around 40–240 villagers depending on its capacity. Based on in-depth interviews with key informants, six illegal sawmills were closed after the implementation of illegal logging inspections in the area.

10.3.2.4 Transportation

The people who transport logs from village to city are also considered to be illegal loggers by the villagers whether they know the log they are transporting is legal or not. Law enforcement officials often catch these people on the way from the village to the city.

10.3.2.5 Head of Illegal Loggers

The people who primarily benefit from illegal logging of forest in Sangir are local Chinese-Indonesians who live in Jakarta. Based on information from key informants, the heads are mainly Chinese and live in Jakarta. The head of illegal loggers influences local people with attractive wages and other incentives and convinces them to engage in illegal logging activities. Sampled respondents believed that only rich people can carry out logging and poor people can only be hired to facilitate logging. However, some key informants argued that it requires only 2 million rupiahs to initiate logging activities, which can be arranged by borrowing from friends. The borrowed money can be returned after selling the timber, harvested through illegal logging, to tauke kayu. The price of wood depends on the quality of the timber.

Illegal logging provides economic benefits for the local households, which motivates them to engage in illegal logging, particularly those with lower educational levels who are unskilled. As shown in Table 10.2 the wage of illegal laborers is enough to support the household compared with the wages of a farmer (Rp 15,000–20,000 per day). Based on in-depth interviews with key informants in the village, and surveys from sampled respondents, since the implementation of illegal logging inspection, the economic condition of the villagers involved in illegal logging has declined.

Illegal logging has some serious environmental consequences. No disaster has occurred in the study area; however, in the neighboring villages erosion occurs once every year and people believe that it is due to logging activities. Valuable species harvested include Red Meranti (*Shorea* spp.), Keruing (*Dipterocarpus* spp.), Mersawa (*Anisopteracostata*), Timbalun, and Banio (*Shorea* spp.). These species are export commodities and have high prices in domestic as well as in international markets. Due to increased illegal logging these species are already dwindling.

10.3.3 Socioeconomic Characteristics of the Community

The population density of the district is around 58 persons per square km. Total population of Lubuk Gadang consists of 18,187 (49.40%) male and 18,628 (50.60%) female, and is increasing at an average growth rate of 5.57% per year. The highest percentage of the population in Lubuk Gadang is dominated by people of productive age (15–56 years) consisting of 13,996 (38.01%) people (Table 10.4).

In Lubuk Gadang there are 18,876 people in the productive age category, where 5% are unemployed. The rate of unemployment in this district, which is around 17% in the district, is influenced by the limited income sources available, the low level of education, and illegal logging inspections, as most of the villagers were involved in illegal logging activities. Thirty percent of the total population in the district are engaged in the agricultural sector while there are also civil servants, traders, and people involved in other occupations in the district (Table 10.5).

TABLE 10.4   Total Population in Lubuk Gadang Based on Age Level

No.	Age Level (Year)	Frequency
1.	0–5	3666
2.	5–7	3828
3.	7–15	7546
4.	15–56	13,996
5.	>56	3444

Source: Sangir on Number, 2005.

TABLE 10.5   Number of People of Productive Age in Lubuk Gadang

No.	Category	Frequency
1.	15–56 years old and working	18,876
2.	15–56 years old and not working	522
3.	15–56 years old (housewife)	5632
4.	15–56 years old and not working (handicapped)	135

Source: Nagari Lubuk Gadang Profile, 2005.

10.3.4 Factors Affecting Illegal Logging

10.3.4.1 Regulation

There are some regulations concerning the forest. These regulations relate to overharvesting of forest products carried out mostly by the private sector and local people involved in forest management.

a. Act No. 5 of 1967 about *Undang-Undang Pokok Kehutanan*.
b. Act No. 33 of 1945.
c. Act No. 41 of 1999 concerning forestry.
d. Chapter 51 GR No. 34 of 2002.
e. President Instruction No. 4 of 2005 concerning illegal logging in all forest areas and its distribution in Indonesia Republic.

Act No. 5 of 1967 about open access to utilize forest product especially at the forest area in outer Java Island. Through Government Regulation No. 21 of 1970 declared all natural resources as state owned, and confined local people's involvement in natural resources management and handed over harvesting rights to HPH companies.

Act 33 of 1945 started recognizing community rights, and involved local people in managing and utilizing forest that is related to community empowerment to enhance social forestry. Social forestry is a program that covers all forest department programs to decrease forest fires, illegal logging, rehabilitation of critical land, structures forest industries, and pushes forest decentralization.

10.3.4.2 Institutions

By eradicating illegal logging the government may face new challenges of unemployment. There is a minimal role of government in controlling illegal logging in the area. Villagers and even outsiders freely exploit timber and other forest products. In some parts of the district it has become a tradition to earn one's living from the forest and forest products. This situation is hard to change as institutions and forestry department officials are also involved in illegal logging activities. Solidarity among villagers is another influencing factor, as almost all parts of the community are engaged in illegal logging.

10.3.4.3 Community

Inspections conducted in the area to tackle illegal logging have made some villagers lose their occupation and become unemployed because they depend on forest and timber for their livelihood. As a result of these inspections and the losses of occupation the economic condition of Sangir declined sharply. This economic downfall of the area is also felt by people who were not involved in illegal logging. For example, the income of a trader or other home industry owner has also declined with the decrease in demand for their products as a result of the financial downfall of consumers.

10.4 CONCLUSIONS AND RECOMMENDATIONS

10.4.1 Conclusions

The main conclusions drawn from the study follow:

1. Illegal logging is an organized activity involving wealthy and powerful people, most of them outsiders. They take advantage of villagers and engage them in illegal logging through attractive wages and other incentives.
2. Socioeconomic characteristics of the community in the study area show that local people have no other choice except to depend on the forest for their livelihoods. The other major occupation in the study area is reported to be farming; however, due to lower wages in farming, people are reluctant to opt for farming as an income source. In this situation, timber harvesting became the most rational decision for local people in the area.
3. The study identified two driving factors for illegal logging: (1) external factors including institutions, markets, and regulations; and (2) internal factors including behavior and knowledge of the community members.

10.4.2 Recommendations

Based on the findings, the following recommendations are forwarded:

1. Recent decentralization and authority devolution to local levels empowers local communities in managing the forest and other natural resources. During field visits it was revealed that local communities are also willing to participate in forest management together with government. There should be a clear mechanism and recognition of communities' right to the use and management of natural resources in the area.
2. Limited employment opportunities induce villagers to engage in illegal logging activities. Government should arrange alternative employment opportunities to discourage illegal logging.
3. Local people must be involved while making laws and rules for forest governance and management. A bottom-up planning process will create a better atmosphere for forest management.

References

AFLEG, 2003. In: Forest Law Enforcement and Governance in Africa: Civil Society Preparatory Process, Yaounde, 13–14 October.
Barber, Charles V., et al. 2002. The State of the Forest Indonesia. World Resources Institute, Indonesia.
Casson, A., Oidzynski, K., 2002. From new order to regional autonomy shifting dynamics of "illegal" logging in Kalimantan, Indonesia. World Dev. 30, 2133–2151.

Casson, A., Setyarso, A., et al., 2005. A Background to Illegal Logging and Law Enforcement in Indonesia and 10 step Program to Tackle Illegal Logging. Draft 13 May, Center for Irrigation, Land and Water Resource and Development Studies.

Krishnaswamy, A., Hanson, A., 1999. Our Forest Our Future: Summary Report of the World Commission on Forest and Sustainable Development. World Commission on Forests and Sustainable Development, Canada.

Mc. Carthy, J.F., 2002. Power and interest on Sumatra's Rainforest frontier: clientelist coalitions, illegal logging and conservation in the Alas Valley. J. Southeast Asian Stud. 33 (1), 77–106.

Siscawati, Mia, 1999. Forest Policy Reform in Indonesia: Has it addressed the Underlying Causes of Deforestation and Forest Degradation? Paper presented at 3rd IGES International Workshop on Forest Conservation Strategies for the Asia and Pacific Region.

WRM, 2005. Illegal Logging Issue No. 98, www.wrm.org.uy/buletin/98/viewpoint-html.

WWF, 2004. Scale of Illegal Logging Around The World: Currently Available Estimates, www.wwf.org.

SOCIOECOLOGICAL SYSTEMS AND NEW FORMS OF GOVERNANCE

Socioecological Aspects of Mandailing Natal People in Buffer Zone of Batang Gadis National Park, North Sumatra: A Case Study on Community in Batahan Village, Enclave Area in Batang Gadis National Park

M.N. *Janra*

Andalas University, Padang, Indonesia

11.1 BACKGROUND

Indonesia possesses high diversity of natural resources. Though it occupies only 1.3% of the Earth, it has around 17% of the world's biotic species. Indonesian forest is occupied by 11% of the world's plants, dwelled in by 12% of all mammals, 15% of all reptiles and amphibians, and 17% of the world's bird species. For instance, in Borneo, there are at least 3000 species of trees, more than 2000 species of orchids, and 1000 fern species (Chidley and Marr, 2002). In Indonesia, tropical forest plays a big role in embracing the economic, social, and cultural aspects of the society (Sumardja, 2002). From 30 to 100 million people of a total of 216 million Indonesian are dependent on the forest; among them, there are approximately 30–50 million indigenous people who are tied to their surrounding forest (Gautam et al., 2000).

Forest dwellers include not only wildlife; human beings have developed their livelihood inside it as well. Along with the improvement of human civilization, forest became an alternative for living. This is the beginning of the term *community forestry*. Community forestry, as explained by Gunter (2004), is a community that involves the three pillars of sustainable development: social, ecological, and economic sustainability. At its core, community forestry is about local control over and enjoyment of the benefits offered by local forest resources. These benefits are both monetary and nonmonetary.

It is in line with the concept of *Forest for People* (a result of The Eighth World Forest Congress in Jakarta in 1978). This concept saw that the biological and socioecological components of forestry issues should hold together, where the people and their forest have a tight bond, composed as an ecosystem. People's behavior can be the measurement of the value of human involvement in natural resources management (Awang, 1999). This concept bears attention and later became the term *socioecology*. Three sets of critical phenomena were outlined: surface terrain and landforms, vegetation communities that occupied those landforms, and land-use practices (settlement, agricultural, freshwater fishery, etc.) (Barton et al., 2006). In the traditional forest community, the aspects of socioecology included traditional customs, landownership and heritage arrangements, arrangements of land clearing to open agricultural area, and grazing pasture. All of these bind the community, while on behalf of the forest community, socioecology enhances their carefulness to maintain the environment.

In North Sumatra, the Conservation International (CI) Indonesia Program is the lead organization for handling conservation efforts. During a Critical Ecosystem Partnership Fund (CEPF) priority refinement process with key partners in the region in early 2002, CI Indonesia learned of a district head (regent who is called a *bupati* in the Indonesian language) interested in setting aside a large tract of forest in his Mandailing Natal District of North Sumatra Province. On Dec. 31, 2003, the regent declared the 108,000 ha Batang Gadis ("Virgin River") a National Park (BGNP) (CBS of Mandailing Natal Regency, 2004).

The central government designated Batang Gadis as a national park on Apr. 29, 2004. Batang Gadis is home to tigers, rhinos, elephants, tapirs, and other globally threatened species, as well as some of the world's highest plant diversity. CI Indonesia is in the process of scientifically documenting and quantifying the park's biodiversity. The park lies at the southern end of the North Sumatra Conservation Corridor and could be the entry point where nongovernmental organizations (NGOs) need to successfully secure protection for large tracts of Sumatra's northern forests. The governor of North Sumatra has pledged to work with CI and its partners to set aside more conservation areas within the greater ecosystem (CBS of Mandailing Natal Regency, 2004).

The BGNP initiative was in accordance with local people's aspirations. Long ago, the Mandailing Natal communities performed a so-called "local awareness" that is still in practice. Traditionally, local communities protect their natural forests and springs, as well as the use natural resources, in wise manner; for example, through the systems of *protected river basin, banua/huta space arrangement, sacred place "naborgo borgo" or "harangan rarangan" (forbidden forest)*, which cannot be disturbed or even damaged. For the Mandailing people, water is the "spring of life," which has a tight bond with social institutions, culture, economics, and ecologic, so its existence must be protected.

Batahan is one of six villages that encompass the Kotanopan Subdistrict and is the most distant village from the capital of the subdistrict town, approximately 20 km west of Kotanopan or 60 km from Panyabungan, the capital town of Mandailing Natal Regency. This village has a definitive legal status and is self-supporting in its characteristics. In a width of 5 ha, it supports about 343 people, with a density level per occupant of 0.81. The population comprises 155 males and 188 females, a sex ratio of 82.50 (CBS of Mandailing Natal Regency, 2004). All of them are Moslems.

As an enclave, Batahan is situated in a valley surrounded by hilly terrain. Villagers' livelihoods are vastly influenced by the existence of *Aek* (River) Batahan, flowing in the southern part of the village and Aek Asollu, which passes through the middle of the settlement and at the end of riffles it meets Aek Batahan and in turn enriches the water mass.

11.2 OBJECTIVES OF THE STUDY

The research is guided by the following objectives:

(a) To document the sociocultural pattern within the local community in Batahan Village, especially in managing their environment that laid in a buffer zone (enclave) of national park.
(b) To analyze the relationship between the sociocultural values of the local people and the ecological performance of the buffer zone of the national park.
(c) To record the factors influencing the utilization of natural resources and environmental services of BGNP to fulfill people's needs

The study is aimed at providing an explanation on natural resources management in an enclave area of BGNP. This typical area, with Mandailing clans as the main actors of natural resources managers inside it, has become an interesting aspect to be explored. The present research study is an attempt to answer the following questions:

(a) How do local people typically manage the local resources that make them different from other forest communities?
(b) To what extent do they perform the best management in their environment?
(c) Why do they practice environmental and natural resource management in their own way?
(d) What are the ecological and social factors that influence the current natural resources management practices?

11.3 METHODS

Primary data on local culture were collected from the Mandailing Natal community in Batahan Village and analyses were performed using a qualitative approach. Techniques adopted for primary data collection included participative observation, independent interviews, focus group discussions, direct observations, and key informants interviews. Secondary data were also used in the study. Main sources of secondary data are documented data available in the Central Bureau of Statistics (CBS) office in the Regency of Mandailing Natal, *Balai Konservasi Sumber Daya Alam* (BKSDA) of North Sumatra, the CI Indonesia Program, and local NGOs (BITRA Konsorsium, 2005).

11.4 RESULT AND DISCUSSION

11.4.1 Landscape Utilization Pattern

Through a horizontal dimension, there are six categories included in the spatial arrangement in Batahan: (1) river and tributary (Aek Batahan and Aek Asollu); (2) rice fields that are situated around the community settlement and grouped into five major regions that reaches around 30 ha in width; (3) settlement; (4) agricultural area, which is called *lodang* or *ladang*; (5) old secondary forest (*harangan rarangan*); and (6) primary forest (*naborgo borgo*), an area that should not be entered (Fig. 11.1). For utilization of its surrounding natural resources, Napitupulu (2006) observed that Mandailings have a concept of spatial arrangement. It consists of five main areas: (1) the settlement area; (2) the fields, ponds, and rivers; (3) the field, pasture, and mixed garden; (4) the forest as a source of medicinal plants, building material, and as a hunting ground; and (5) the forbidden forest, where people cannot enter and is believed to have magical properties.

11.4.2 Agricultural Sectors

Villagers have converted around 30 ha of forestland around the village and some parts in the other side of Aek Batahan into cultivation area, mainly for rice. There are five main locations for rice cultivation: *Saba* (rice field) Bondar Lubuk Gala-Gala, *Saba* Aek Damar, *Saba* Muara Tombang, *Saba* Tano Barani, and *Saba* Aek Asollu. These 30 ha of paddy field were divided into 20 ha in the upstream area managed by 58 families and the rest in the downstream area cultivated by 12 farming families.

Agroforestry is a mixture of components that consist of woody plants (timber, clump, palm, bamboo, and other cambium-borne plant species) with agricultural plants (seasonal species) and/or cattle, which set in temporal a arrangement and spatial arrangement as well (Sardjono et al., 2003). The best possible way to develop an agroforestry system in the area is by using forest resources (in the form of forest-based agroforestry), started through clearing wood/shrub and plowing and cultivation activities.

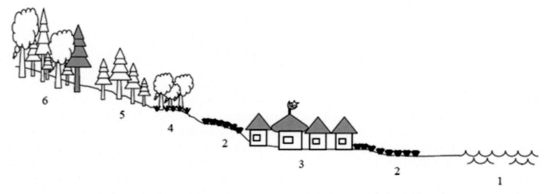

FIG. 11.1   Batahan landscape in horizontal view. 1. River, Aek Batahan; 2. Rice fields; 3. Settlement; 4. Agricultural area (*lodang*); 5. Old secondary forest (*harangan rarangan*); 6. Primary forest (*naborgo borgo*).

There are local varieties of rice in Batahan that have been cultivated for a long time. People in Batahan call these varieties *Silatihan* and *Sigodang*. The first one is preferred more by the villagers than the latter one. The total growth period of these varieties is around 7 months; hence, villagers have only one planting time a year starting in the 11th month of the Hijriah calendar.

Irrigation channels are repaired during the month of *Dzulqai'dah* while hoeing activities are carried out in Muharram (New Year of the Hijriah calendar). Replanting of seedlings into land is carried out in Safar. Harvesting of the crop takes place in Dzulqai'dah month in the next year. During the spare time (from transplantation to harvesting) local people are engaged in handling their *lading* (to rejuvenate aromatic oil plants and distil the mature ones to extract the oil content). The whole arrangement of the planting and harvest period can be seen in Fig. 11.2.

Ladang in Batahan is a complex mixture of various productive plant species, divided into several categories (habits, benefits, and functions). A short observation within a few *ladang* locations in Batahan, had listed more than 40 useful plant species that exist in a single hectare of land (see Table 11.1). They are all grown in an overlapping manner to each other, but created conditions similar to a natural forest; consisting of upper, middle, and lower layers. *Ladang* as an agricultural system in the tropical area that evolved more than 7000 years ago. It included pastureland, old garden, and even old secondary forest, converted into a human-made ecosystem. Land clearing was pursued by a cutting and burning technique without efforts to maintain soil fertility. After the runoff of soil nutrition and/or the emergence of weeds, humans usually leave the field and start clearing other forest areas for agricultural purposes.

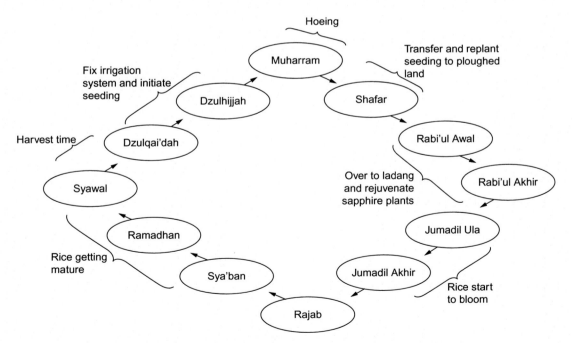

FIG. 11.2 Arrangement of planting periods in Batahan Village, using the Arabic calendar system.

TABLE 11.1   List of Plant Species Cultivated in *Ladang* in Batahan Village

No.	Vern Name	Habit	Contributions
1.	Manggis (*Garcinia mangostana*)	Tree	Consumption, fuel wood
2.	Durian (*Durio zibethinus*)	Tree	Consumption, fuel wood
3.	Jeruk (*Citrus* sp.)	Tree	Consumption
4.	Aren (*Arenga pinnata*)	Tree	Consumption, building material
5.	Bambu-Betung (*Bambusa* sp.)	Tree	Consumption, building material
6.	Pisang (*Musa paradisica*)	Tree	Consumption
7.	Tebu (*Saccharum officinarum*)	Pole	Sugar ingredient
8.	Cengkeh	Tree	Spice, commercial commodity
9.	Kelapa (*Cocos nucifera*)	Tree	Consumption
10.	Jambu Bol (*Psidium jambos*)	Tree	Consumption
11.	Pepaya (*Carica papaya*)	Pole	Consumption
12.	Duku	Tree	Consumption
13.	Kayu manis (*Cinnamomum burmannii*)	Tree	Commercial commodity
14.	Karet (*Hevea brasiliensis*)	Tree	Commercial commodity
15.	Jengkol (*Pithecellobium lobatum*)	Tree	Consumption
16.	Kemiri	Tree	Consumption, spice, commercial commodity
17.	Kopi (*Coffea robusta*)	Tree	Commercial commodity
18.	Rambai	Tree	Consumption
19.	Salak (*Salacca edulis*)	Tree	Consumption
20.	Mangga (*Mangifera indica*)	Tree	Consumption
21.	Jambu air (*Psidium aquea*)	Tree	Consumption
22.	Jambu biji (*Psidium guajava*)	Tree	Consumption
23.	Petai (*Parkia speciosa*)	Tree	Consumption
24.	Pinang (*Areca catechu*)	Tree	Consumption, commercial commodity
25.	Cempedak (*Artocarpus integra*)	Tree	Consumption
26.	Petai China	Tree	Consumption
27.	Ubi jalar	Herb	Consumption
28.	Buncis	Herb	Consumption
29.	Ubi kayu (*Manihot utilissima*)	Shrub	Consumption
30.	Labu	Herb	Consumption
31.	Labu siam	Herb	Consumption

TABLE 11.1   List of Plant Species Cultivated in *Ladang* in Batahan Village—cont'd

No.	Vern Name	Habit	Contributions
32.	Nilam	Shrub	Commercial commodity
33.	Cabai merah (*Capsicum annuum*)	Shrub	Consumption
34.	Terong	Shrub	Consumption
35.	Cabe rawit (*Capsicum frutescens*)	Shrub	Consumption
36.	Jahe	Herb	Consumption

Source: Primary Data, 2006.

This system, according to Kontjaraningrat (1984, in Yantodium, 2006), needs a considerable width of land and even takes into account some part of the virgin forest.

The agricultural sector in Indonesia is generally characterized by transitional subsistence farming, not commercial agriculture (unless possessed by the government or the private sector). It may be more familiar as a communal agricultural system, marked by a considerably narrow area, simple technology, intensive labor capital, and a mixture of planted crops species (Hernanto, 1993). It was also observed that the people in Batahan converted the southern part of the village, which was considered as the most fertile area, into agricultural land. Soil fertility can be categorized as good, because the harvest system does not involve total land clearing (like cinnamon), but instead proceeds in a slow and multistage manner. The owners of the *ladang* cultivates crops according to their needs, with commodities that match market demand, emergency needs, and so on. This enables the land to convalesce naturally and guarantees the availability of soil nutrition through the leaf litter produced.

11.4.3 Preserved Forest (*Harangan Rarangan*)

Harangan rarangan is a term that refers to forest patches around the village (mainly in the upstream region) that are not allowed for conversion by any person, unless under customary legalization. These areas are marked by several spots, which show the remnants of agricultural activities in the past. There are leftover tree stands of medium size, consisting of cinnamon, mahogany, and others, signs of the interference of humans in the natural ecosystem. Parts of forest patches in northern and western Batahan are springs areas of two rivers included in the *harangan rarangan*. The spring area of Aek Asollu had even been dashed against a landslide, which made villagers around this area move temporarily to the western part of the village.

Most of the forest in Batahan is not excluded from the *harangan rarangan* that is supposed to avoid deforestation. If a villager wants to convert a forest patch into agricultural land (which is not defined in the *harangan rarangan*), he should make a contract with the *hotabangon* (custom headman). There is an obligation to replant any space that has usually been inert after land clearing, which precedes every cultivation process. Customary fines and sanctions are the consequence for any collusion. People usually do not have any objection if an outsider wants to plow some part of the land, but they must ask permission from the village chief and the custom headman; thus, it can be decided based on the village chief's decision.

Batahan people always consider every natural resource in and around them as communal belongings, except for those in which they personally labored. As long as there are certain remnants in a particular area, collective agreement is conceived by others. The area for a specific cultivator is marked with a tree, such as *dadap*, cinnamon, and mahogany. However, this does not apply to hunting of animals in the forest, where an animal belongs to the person who first sees and catches it. Communal resources are resources that are not the personal property of someone; instead, all people have the same rights to use the resources. Communal resources tend to deplete more rapidly than other resources.

According to Sumardi (1997, in Syahrawati, 2004) villagers still hold tight to custom norms and traditional heritage, which is passed through from one generation to the next. Though the level of dependency of indigenous people on forest resources is high, the norms and traditions will not let them exploit it in an unsustainable manner, even for commercial purposes. There are many examples where forest communities appropriately manage the environment, whether it is derived from pure traditional norms, or delegated through the mixture of traditional and governmental policy. This can be found in the *Dayak* community of Kalimantan where villagers have arranged and utilized their simple but efficient traditional technology to cultivate various species of rattan. They also mastered the technology in benefit taking, harvesting, protection, and marketing those things that are considerably harmless to the forest resources (Djuwantoko, 2000, in Syahrawati, 2004).

11.4.4 Sacred Forest (*Naborgo Borgo*)

Huta are woody patches that are believed to be the home of invisible creatures. People in Batahan reported that these creatures have same social life as human beings. They live in a somewhat penetrable forest (primary forest). Various stories tried to reveal the beginning of this myth. The most famous version is a story about a goat shepherd who had embarrassed himself in front of people. He wanted to fulfill his promise to slaughter his best animal and serve the people who are led by a pious clergy. Nevertheless, he denied what he promised before; just pick the feeblest and smallest goat to be served in a dish. To substitute the insufficiency, he slaughtered a dog (*kanjing*) and a cat. He prepared food from the flesh of the goat, dog, and cat separately into three piles accompanied by the barking, meow, and blare sound of each animal. Villagers soon realized and were disappointed because of his act. The guilty person felt embarrassed, ran away from his village, and entered the deep jungle. He whispered to God to let him live in his embarrassment and make him detach from the community. Since then he turned into an invisible creature, forming his own social life and was termed *naborgo borgo*.

This story is accepted on faith from one generation to the next in Batahan, as well in surrounding villages. Nowadays, people in Batahan realize their faith through an annual vow (*nazar tahunan*) in Hajj month (month in the Islamic calendar) as a preventive way to any contiguity between humans and this invisible creature. Although it is believed that no dispute ever occurred, people usually avoid entering this closed primary forest (*naborgo borgo*), which created a form of conservation to forest resources in Batahan Village and the surrounding area.

In relation to the religiosity concern, forest is considered as a place, which possesses sacred or magical property, and significantly affects the cultural system. It is widely believed that the forest is not only a physical environment, but also a structure that contains and is filled by spirit, and will react and be damaged instantly if it is treated in an unsustainable manner.

In response to concern over the high rate of land clearing and forest exploitation, village elders in Keluru, Kerinci Regency had established *Rimbo* (forest) Temedak as a region where their *Hukum Adat* (traditional law) has been applied. The mutual agreement among the elders in that place has produced 11 rules, consisting of protecting *Rimbo* Temedak, containing some prohibitions, rights and obligations, fines, and replanting cost for any forest damage caused by human action (Nuansa Lingkungan, 2000, in Syahrawati, 2004). This fact shows that the customary community can be trusted to handle the management of natural resources in their own environment. The traditional forest community in Bengkulu is another example of community forestry where villagers have divided the forest into several categories including prohibited forest or *Imbo Lem* (dark forest).

11.4.5 Pattern in Water Resource Utilization

11.4.5.1 River

Almost all of the resident areas in Batahan are supported by two rivers across the village; namely, Aek Batahan and Aek Asollu, where the first is the main source for irrigation water in agriculture and fishing activity and the latter for household consumption and sanitation. Both rivers have a balanced portion for fulfillment of daily needs. Aek Batahan has an upstream region in *Huta Nagodang* (the large forest) and Aek Asollu which pates in Tor Asollu, next to the village. Sustainability of water resource is supported by the other part of the village landscape, which takes the forest (primary and secondary) and *ladang* into account. The incidences of landslides in the past induced villagers in Batahan to maintain the forest. Logging in the forest at the upstream area (spring) is a breach of customary provisions and considered as violence. It has been a heritage from an older generation, which is in line with Steward (1979, in Eldiny, 2005) who argued that culture and knowledge possessed by a certain community and used to interact with the environment through norms must contain a role to behave in a certain manner toward nature and a role learned through symbolic communication and that has emerged through the conclusion based on experience.

Therefore, people in Batahan have created roles concerning their two main rivers across the village to sustain the availability of water and prevent any disaster that will possibly impact them as a result of violence of humans. The role of those two rivers can be viewed in Fig. 11.3. Both Aek Asollu and Aek Batahan play a similar role in irrigation as their position is next to the location of five main rice field areas. A vast area of rice field exists around the settlement being irrigated by Aek Asollu; meanwhile, the rest depend on Aek Batahan. In other words, Aek Asollu is more favorable for household consumption and sanitation, as it crosses directly the settlement area, enabling each villager to access freshwater easily. Aek Batahan is mainly used more for bathing than for washing and drinking. For fishery, there is a significant difference between Aek Asollu and Aek Batahan. The first flows into artificial pools around village, which are used to raise several species of fishes (*Cyprinus* spp., *Tilapia* spp.). The latter gives major fishery yield, gained through nets, fishing rods, *lukah,* and other catching tools. Aek Asollu holds three main functions: (1) powering a traditional rice mill (*losung bonggar*); (2) religious services because it flows side by side to the main mosque and is also channeled to each corner of the village where small mosques (*mushola*) are located; and (3) for processing

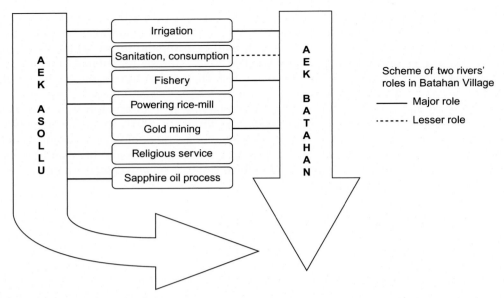

FIG. 11.3   Role of rivers in Batahan Village.

sapphire oil. Aek Batahan also possesses the same capacity for roles one and three; however, because of the difficulty to control water flow in the rainy season, people choose Aek Asollu where the flow is comparatively stable.

11.4.5.2 Up-to-Downstream Utilization and Management

Keeping in view the importance of water availability to support life in the village, there is an integration of water resource benefits taking place in Batahan. Management in the exploitation of water resources must pay attention to the spatial arrangement of landscape. It must follow certain roles, facilitate the process of avoiding erosion, and reduction in water flow that can result in degradation of land fertility and poses potential threats of natural disaster. There should also be a restriction against water pollution, whether it is in rivers, lakes, or marshes (Lubis, 2006). Both utilization and management efforts of the two Batahan rivers can be seen in Fig. 11.4.

The length of each river and its capacity for serving the village's needs determine the form of human's interaction toward it. Aek Batahan has more functions in the community than Aek Asollu. However, both of them have the same treatment in the upstream area and are protected under *harangan rarangan* to prevent water shortages. *Harangan rarangan* together with *naborgo borgo* had become institutional tools to define a set of rules that bind, arrange, and bear attention from all members of the community. It again defines the cooperation management and coordination among people in the community to use natural resources (Kasryno, 1984, in Yusmini, 1994). The water bodies are mostly free of chemical substances as the use of chemical fertilizers and pesticides in agriculture is low in the area. Though there are no rules on the application of chemical fertilizers and pesticides in agriculture, higher market prices discourage its use in agriculture and result in chemical-free water in the rivers.

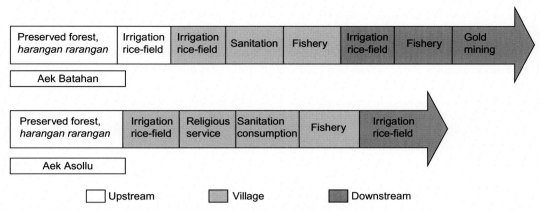

FIG. 11.4   Utilization and management efforts of rivers in Batahan Village.

11.4.5.3 River Border

As a part of the spatial arrangement concerning the existence of water bodies, people have applied certain rules for maintaining and minimizing destruction originating from river flow. Although the settlement is located in a secure location from Aek Batahan (the most possible for any waterborne disaster), some rice field generally is laid at the edge of the river. Therefore, there is an unwritten rule that farmers must put a causeway for rice field in such a space to the river's edge, which usually ranges from 2 to 5 m. Even in the stable flow, there is still potential for riverbanks erosion. Situating a considerable distance between rice fields (or *ladang*) to riverbanks is a safe way to restrain any possibility of this issue in the future.

Syahrawati (2004) reported the customary laws on river water utilization in Alam Surambi, Sungai Pagu, and West Sumatra that encompasses three rules: (1) Do not make turbid water flow, (2) Do not take river stone in a destructive manner; and (3) Do not hamper the riverbank.

There will be an innate sanction for any breach where the guilty person must slaughter cattle. A ritual ceremony would be held in the upstream area, where the elders of the village advises the whole community and reminds them of the custom law.

11.4.5.4 Protection in Spring Spots

A steady rule has been stated by the elders in Batahan to maintain the green cover upstream. Aek Asollu has upstream area in Tor Asollu, near the village and Aek Batahan, which pates in *Huta Nagodang*, and both of those become the *harangan rarangan* (BITRA Consortium, 2005). Tree logging in these two places is forbidden. Any violation will result in heavy penalization. This customary rule is endorsed and embraced by the younger generation. It has been proved that all age group individuals in Batahan (youths, adults, and elders) agreed to protect their *harangan rarangan* in Tor Asollu and Huta Nagodang based on their understanding of customary values. Each of them has accepted same experience through the elders' story concerning nature behavior following human destruction efforts. Cultural values have more suppressive power, because of their innate characteristics that are derived from actual reality. Lessons from nature, in the form of disasters, always create thought on how to avoid the same event in the future. The result of the effort of thinking usually endows the younger generation, who assume learning from ancestors is a truth (Steward, 1979, in Eldiny, 2005).

11.4.6 Local Wisdom and Traditional Value in Regard to Natural Resources Management

Long-term interaction between man and nature had created a one-way dependency of humans on nature. Adaptation is a tool for living in an environment, where each kind of habitat establishes a typical culture that differs from others. In line with the laws of nature, the Earth moves in a very regular and well-arranged manner, creates season, wind, weather, and the changes between night and day in a consequent manner. To survive, humans have to evolve their mind to record any nature signals, which are repeated regularly. Understanding nature's behavior is called *local wisdom,* which is transferred from elders to youngsters and is an important aspect to build a solid civilization.

11.4.6.1 Fuel Wood Extraction

Fuel wood is the main source of energy for cooking, heating houses, and the preservation of wooden material. Origin of fuel wood in Batahan is derived primarily from fallen timbers and dead trees, which are available in abundance within the forest and the *ladang*.

11.4.6.2 Prohibition Against Poison and Electrical Snatchers in Catching Fish

Fishermen have made a sturdy rule regarding harvesting fish from Aek Batahan, prohibiting the use of poison and electrical snatchers as tools to catch fish. A person violating this rule will be fined around two and half million rupiah.

11.4.6.3 Natural Phenomena

Villagers in Batahan consider various natural phenomena as signs of starting new activities or adjusting their existing activities. Based on information from key informants, there are several natural occurrences still believed to be true. They are as follows:

(a) If a clutch of wild bees fly away to the east of the village, it is a sign that the dry season is coming.
(b) If a bird called a *Sikakap* sounds in the middle of the night, it is an augury for bad news where somebody will possibly die.
(c) If a butterfly enters the house, it means the arrival of a guest. If it is a moth, it is an adverse sign, meaning one of the family members will pass away.
(d) If a tapir (*sipan*) breaks down into the edge of the village, villagers interpret this event as a warning to not enter the jungle (both *harangan rarangan* or *naborgo borgo*). A tapir will not enter the settlement unless followed by another wild animal, which is usually a tiger.
(e) If a tiger roars around the village, it is a sign there has been adultery among the people in the village. Some violation like any sexual act without marriage or premarital courtship (between a man and a woman) will invite the tiger into the village. This sign is also believed in other places in Madina Regency and other parts of Sumatra, such as among the Minangkabau people of West Sumatra, where there is a superstition that says tigers are the reincarnation of an ancient spirit, so they will watch their descents in the village, and give warnings for any violation that happened.
(f) When the whirlwind flows for a long time (called as *tarutung*), it is usually followed by the blooming period of durian and several fruit plants.

11.4.7 Stakeholders in Batahan Village

As the lowest unit of governmental strata, the Batahan community also has a set of village authorities that consist of a village headman, vice of village headman, and a village secretary. They seek guidance from the *Badan Permusyawaratan Desa* (BPD) (Village Assembly Council) while making decisions regarding problems faced by villagers. Youth and women groups are rare in the village, possibly because women seem less interested in being involved in formal cooperation activities; instead, they are engaged in the rice field or *lading* and in caring for their children and other family members. Their work burden forces them not to be involved in such groups or cooperation.

Another significant formal institution in the village is a NGO, which is termed *Organisasi Konservasi Rakyat* (OKR) (People's Conservation Organization). This NGO was established through the enhancement of the BITRA Consortium in response to the initiation of the BGNP. Through counseling from the BITRA Consortium, there are already 35 villages (out of 71 villages) around BGNP (most of them are included in the buffer zone of this national park) having similar institutions. In Batahan, OKR has role to facilitate each training session held by BITRA Consortium, especially in agricultural and gardening sectors. The current activities of the OKR include development of alternative income sources for the community to raise the living standard of villagers.

Along with formal leaders, villagers also possess informal figures, known as *hatobangon*. This custom figure is responsible to settle any problems inside the village, including disputes, marital issues, custom roles, applying fines, banishment, and so on. Other influential parties include wealthy traders, the committee of *madrasah*, and mosque committees.

11.5 CONCLUSION

Batahan villagers already have some long-established customary values (*harangan rarangan, naborgo borgo,* and *lubuk larangan*) and social norms (*forbidden, natural phenomena*) for interacting with the environment, which significantly regulates their efforts to take processes and treat natural resources. These norms and values were the products of a long-term interaction between villagers and their environment, transferred from one generation to the next.

Observation of the natural and environmental condition in Batahan revealed that the forest and wood patches around the village are considerably good. Several indicators that were used to conclude this were the presence of key animal species, which are usually used for habitat quality assessment, such as birds (Hornbill, Treepie, Laughingthrush) (MacKinnon et al., 1998) and mammals (tigers, tapirs, and honey bears) (see Table 11.2).

The establishment of sociocultural values gives great hope for conservation of the natural resources base. Long-time interaction with the environment made the villagers realize the importance of natural resources and the benefits of obeying social norms and cultural values that are held in the highest position by all villagers. There are some other factors facilitating the sustainable utilization of natural resources in the area. These are listed as

TABLE 11.2   Bird Species Observed Around Batahan Village Using the MacKinnon Method

1.	*Phaenicophaeus chlorophaea* (Raffles' Malkoha)
2.	*Collocalia esculenta* (Glossy Swiftlet)
3.	*Psilopogon pyrolophus* (Fire-tufted Barbet)
4.	*Blythipicus rubiginosus* (Maroon Woodpecker)
5.	*Buceros rhinoceros* (Rhinoceros Hornbill)
6.	*Chloropsis cyanopogon* (Lesser Green Leafbird)
7.	*Chloropsis sonnerati* (Greater Green Leafbird)
8.	*Chloropsis venusta* (Blue-masked Leafbird)[a]
9.	*Pycnonotus goiavier* (Yellow-vented Bulbul)
10.	*Eurylaimus ochromalus* (Black-and-yellow Broadbill)
11.	*Stachyris* sp. (Babbler)
12.	*Dendrocitta occipitalis* (Sumatran Treepie)[a]
13.	*Dicrurus leucophaeus* (Ashy Drongo)
14.	*Copsychus saularis* (Magpie Robin)
15.	*Enicurus velatus* (Lesser Forktail)
16.	*Garrulax palliatus* (Sunda Laughingthrush)[b]
17.	*Garrulax leucolophus* (White-crested Laughingthrush)[c]
18.	*Artamus leucorynchus* (White-breasted Wood-swallow)
19.	*Prinia familiaris* (Bar-winged Prinia)
20.	*Orthotomus ruficeps* (Ashy Tailordbird)
21.	*Anthreptes malacensis* (Plain-throated Sunbird)
22.	*Passer montanus* (Eurasian Tree Sparrow)
23.	*Lonchura punctulata* (Scaly-breasted Munia)

[a] *Sumatran endemic species.*
[b] *Endemic in Sumatra and Borneo.*
[c] *Asian species, in Indonesia only found in Sumatra (according to MacKinnon et al., 1998).*
Source: Primary Data, 2006.

(a) Believe in the existence of the ancient spirit (ie, *naborgo borgo* in Batahan).
(b) The existence of *Lembaga Adat* (customary council), which has become the source of rules in the village. The council establishes a public statement on environmental utilization and management (with rules empowerment and sanction application inserted)
(c) A central customary figure (*hutabolang*) plays a role in establishing and implementing *adat* rule, supervising *adat* property, and imposing sanctions on every breach of *adat* rule.

Acknowledgments

The author appreciates the immense support from Dr. Ardinis Arba'in and Sukanda Husin, LLM, as supervisors in giving advice during research and manuscript preparation; profound gratitude to Dr. Rudy Febriamansyah and Dr. Helmi for their continuous support; and Riswan Saleh Siregar as the connector in language translation during fieldwork in Batahan Village. We also want to deliver great thanks for assistance during the fieldwork session from Hendra, Enda Mora (BITRA Consortium, Penyabungan), Pak Nasruddin, and Bang Ramlan as part of the Batahan community that contributed a lot to this study.

References

Awang, S.A. (Ed.), 1999. Forest for People Berbasis Ekosistem (Pustaka Hutan Rakyat). BIGRAF Publishing, Yogyakarta.

Barton, M., Sarjoughian, H., Falconer, S., Mitasova, H., Arrowsmith, R., Fall, P., 2006. Modeling long-term landscape dynamics and the emergence of intensification. In: Invited Symposium Paper Presented at the 71st Annual Meeting of the Society for American Archaeology, San Juan.

BITRA Konsorsium, 2005. Dari Hutan Rarangan ke Taman Nasional. Potret Komunitas Lokal di Sekitar Taman Nasional Batang Gadis. USU Press, Sumatra Utara (in Bahasa).

Central Board of Statistic (CBS) of Mandailing Natal Regency, 2004. Kotanopan sub district in number of 2004. Penyabungan. Conservation international Indonesia programme. Critical Ecosystem Partnership Fund (CEPF) Report, June 23–25, 2005.

Chidley, L., Marr, C. (Eds.), 2002. Forests People and Rights: A Down to Earth Special Report, June 2002. Forest People Programme—Rainforest Foundation http://dte.gn.apc.org/srfi1.pdf (accessed 05.02.06).

Eldiny, L., 2005. Tingkat Peran Serta Masyarakat dalam Pelestarian Lingkungan dan Faktor-Faktor yang Mempengaruhinya pada Taman Nasional Bukit Tiga Puluh (Kasus Desa Rantau Langsat Kabupaten Indragiri Hulu Provinsi Riau). Program Pasca Sarjana, Universitas Andalas, Padang. Unpublished.

Gautam, M., Lele, U., Kartodihardjo, H., Khan, A., Erwinsyah, Rana, S., 2000. Indonesia—The Challenges of World Bank Involvement in Forests. OED Report of World Bank, World Bank, Washington, DC.

Gunter, J. (Ed.), 2004. The Community Forestry Guidebook: Tools and Techniques for Communities in British Columbia. FORREX-Forest Research Extension Partnership/British Columbia Community Forest Association, Kamloops/Kaslo, BC. FORREX Series Report No. 15.

Hernanto, F., 1993. Ilmu Usaha Tani. Penebar Swadaya, Jakarta (in Bahasa).

Lubis, R.F., 2006. Air Sebagai Parameter Kendali Dalam Tata Ruang (in Bahasa), http://io.ppi-jepang.org/article.php?id=171 (accessed 08.08.06).

MacKinnon, J., Phillipps, K., Balen, B.v., 1998. Panduan Lapangan Pengenalan Jenis-Jenis Burung di Sumatera, Kalimantan, Jawa, Bali dan Kalimantan. Birdlife International—Indonesia Programme, Puslitbang Biologi LIPI, Bogor (in Bahasa).

Napitupulu, L., 2006. An Assessment of Opportunities for Intervention in Mandailing Natal and the Gayo Highland of Northern Sumatra. Conservation International Indonesia, Jakarta.

Sardjono, M.A., Djogo, T., Arifin, H.S., Wijayanto, N., 2003. Klasifikasi dan Pola Kombinasi Komponen Agroforestri. Bahan Ajaran 2, World Agroforestry Center (ICRAF), Bogor (in Bahasa).

Sumardja, E.A., 2002. Pemanfaatan Sumber Hayati Secara Berkelanjutan. Disampaikan pada Seminar Aktifitas Penelitian Program Konservasi Keanekaragaman Hayati. Kerjasama Antara Pusat Penelitian Biologi LIPI (in Bahasa).

Syahrawati, M., 2004. Menggali kearifan lokal Masyarakat Alam Surambi Sungai Pagu terhadap sumberdaya hutan. In Jurnal VISI 26, 35–36. Padang: Pusat Studi Irigasi, Universitas Andalas (in Bahasa).

Yantodium, 2006. Kelestarian Taman Nasional Kerinci Seblat dan Kaitannya dengan Kondisi Sosial Ekonomi Petani Peladang di Taman Nasional Kerinci Seblat (Studi Kasus di Kecamatan Gunung Kerinci). Program Pasca Sarjana, Universitas Andalas, Padang. Unpublished.

Yusmini, 1994. Dampak Pembangunan Irigasi Terhadap Kelembagaan Penguasaan Lahan (Land tenur) di Kecamatan Baso, Kabupaten Agam Sumatra Barat. Fakultas Pasca Sarjana KPK IPB-UNAND, Bogor.

CHAPTER

12

Hydrologic Characteristics, Flood Occurrence, and Community Preparedness in Coping With Floods at Air Dingin Watershed, Padang, West Sumatra

Yenni, Helmi, Hermansah

Andalas University, Padang, Indonesia

12.1 INTRODUCTION

The Air Dingin watershed is situated at the confluence of the Air Dingin River, a resource for drainage channel, drinking water, livelihood activities, and also for final disposal of wastewater and solid waste (Abuzar, 2005). It is located in Koto Tangah Subdistrict, Padang, West Sumatra, and covers an area as much as 145.57 km², which lies on latitude 0°43′44″–0°54′0″ and longitude 100°19′38″–100°30′45″ (BPS, 2006). More than 99% of settlements and development areas take place in the lowland (BPS, 2005).

Air Dingin region is constituted by 11 villages, the residence for 119,043 persons or about 28,683 households. Each household consists of 4–5 members. There is an equal sex ratio between males and females. None is more dominant than others. Population density in each village varies from 301 to 3760 persons/km² (BPS, 2006).

The socioeconomic and demographic characteristics of the sampled respondents indicates that the occupations of inhabitants vary widely. They are farmers, civil servants, and traders. They grow paddy, coconut, and seasonal crops. They raise goats, carabaos, and buffaloes. Fourteen percent of residents are classified as poor households. Most of residents are 0–54 years old. Only 0.065% of inhabitants are above 54 years old.

Before structural development of the Air Dingin River was put into operation in 1998, this region was subjected to floods every monsoon season. Extreme flow usually occurs in

September, October, November, and December, and sometimes continues until January. The highest flow usually comes in November when rainfall exceeds 100 mm/24 h. The inhabitants suffer from 2 to 5 days of flood with 1.5–2 m elevation on average. The Water Resources Agency records that huge and devastating floods occurred in the years 1972, 1979, 1980, 1981, 1982, and 1986, which caused loss of life and property. Distribution of the flood risk area in the Air Dingin watershed is shown in Fig. 12.1. It covers about 37% of the total flood risk area of Padang.

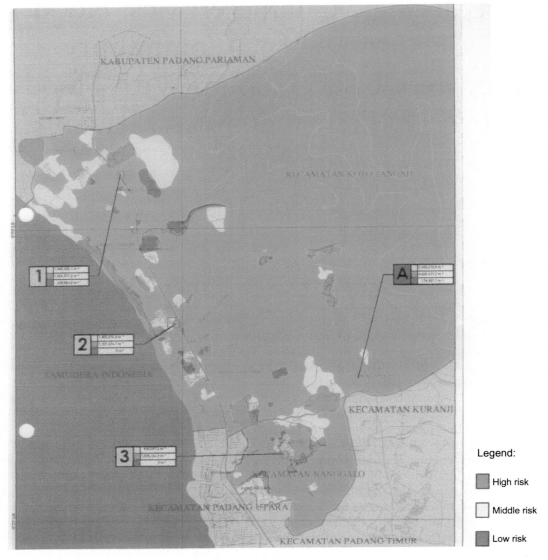

FIG. 12.1   Distribution of the flood risk area in the Air Dingin watershed.

The historic reason for floods in the Air Dingin watershed is that rising water levels in the river inundate adjacent areas due to the watershed's shape. The shape is like a bird's feather, narrow and long, extending from the upland area to the estuary, which means a long time is required for the surface flow to reach the ocean. Huge and continuous rainfall needs to flow along this tiny and long watershed creating high peak flow because the upper reaches of the watershed will be contributing runoff to the peak at the same time as the storm is over the outlet of the watershed, enlarging the flood potential. Moreover, the river path is curved in some places, blocking the river flow. On the other hand, high tides block the flow into the ocean, creating a water hammer to the inland. This situation deteriorates further from the addition of dumped solid waste along the river.

Structural development in the Air Dingin watershed was originally set up in 1997. This measure consists of normalization of the Air Dingin River to increase its capacity and the development of a dike. It was designed by the Japanese consultant, Nikken Consultant, in 1983 using limited data on the hydrologic characteristics of the watershed. As consequence, it was based on limited consideration on flood characteristics in the Air Dingin watershed. Even after structural development was completely finished in the year 2000, the region still received 50 cm of periodic monsoon flooding (shallow flooding) in most monsoon months.

An important lesson learned from others is that completion of structural measures alone cannot guarantee safety from flood loss to any region. There will always be possibilities of failure to any structural development. A sustainable, integrated floodplain management plan that relies on community capacities with the help of science, and support from the government and other stakeholders, needs to be developed. This chapter specifically examines and explores options to improve flood mitigation strategies for the Air Dingin watershed.

The study has been carried out in a number of steps, which include reviewing the existing documents on floods, field visits, exploring hydrologic characteristics of the Air Dingin watershed, interviews with inhabitants in flood prone areas and responsible governmental agencies, developing a flood forecasting model using PCRaster, calibrating the flood forecasting model, analysis of the flood forecasting model, reviewing current flood management practices by government agencies, reviewing current strategies of the community in coping with floods, and the synthesis of an integrated community-based flood forecasting and early warning mechanism.

To measure the success of the research, there is a need to formulate the objectives of the study. They are

 i. To assess hydrologic characteristics of the Air Dingin watershed, Padang City.
 ii. To analyze flood occurrence in this watershed.
 iii. To assess community preparedness in facing flooding in this watershed.
 iv. To develop an integrated community-based flood forecasting and early warning system.

12.2 METHODOLOGY OF RESEARCH

This study develops a flood forecasting and early warning system for the Air Dingin watershed based on quantitative and qualitative methods. The quantitative method is used in developing the flood forecasting model using PCRaster software. The qualitative method is used in developing the community-based flood early warning system. The general logic of thinking of this study is described in Fig. 12.2.

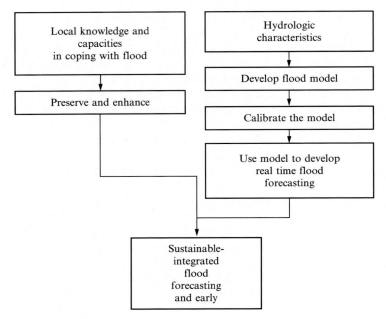

FIG. 12.2   General logic of thinking.

12.3 RESULTS AND DISCUSSIONS

12.3.1 Hydrologic Characteristics of the Air Dingin Watershed

12.3.1.1 Digital Elevation Model

Digital terrain data are topographic data, including elevations, slope, and so on. The interpolation of a derived point into an image surface via spatial interpolation is a crucial process to build a digital elevation model (DEM). A DEM is a function of elevation with respect to a geographic point, $Z = f(X,Y)$ (Aziz, 2001).

Data on a DEM was gained from Gasitech Consultant. A DEM was compiled in the year 2000. The topographic map was derived from photogrammetric techniques (50–100 m spacing) and consists of contour lines. The map includes a geographic (latitude/longitude) reference, Universal Transverse Mercator, World Geodetic System 1984 datum Cent. Meridian 99d E. The Air Dingin region has altitude ranging from 0 to 1860 m above mean sea level. The highest place is at the upland area positioned in the east part of the region. The altitude decreases gradually to the west. There is no extreme variation in slope. The catchment consists of several subcatchments formed by a subriver with a river length of 26.50 km. Subrivers in the upland have steeper gradients than most of the catchment.

12.3.1.2 Precipitation

There is no extreme temperature variation between daylight and night. Temperature ranges from 23°C to 32°C in the daylight and decrease to 22–28°C at night. Average rainfall is

6000 mm/year. Twenty-two years of daily rainfall records from 1979 to 2001 at the four nearest rain gauge stations were collected from the water resource agency and climatology stations. The stations are Tabing (S 0°53′00″/E 100°22′00″), Gunung Sagrik (S 00°53′02″/E 100°24′24″), Kasang II (S 00°53′28″/E 100°25′07″), and Simpang Alai (S 00°56′04″/E 100°26′20″). Blank records were completed by the ratio normal method. Complete data were then checked for quality and consistency by using the double-mass curve technique. Correction will produce consistent rainfall data for the next analysis.

Locations of rain gauge stations were stored to the base map of Padang City using AutoCad Land Development software. The resulting map was exported into ArcGIS software. Here, distribution of rainfall was determined based on real coordinates of rain gauge stations using the Thiessen polygon method. The map was saved in raster form and converted into an ASCII file to be analyzed using PCRaster GIS analysis software. A dynamic model was developed using 30 records of daily rainfall data.

12.3.1.3 Infiltration

Data on infiltration were measured through a field survey using double-ring infiltrometer technique. Sampling points were determined based on a map of soil type produced by Agam Kuantan Watershed Management Body (BPDAS Agam Kuantan). The terrain consists of latosol, podzolic, regosol, litosol, organosol, alluvial, and gleysol. There is no detailed information on spatial distribution of these soil types. The map classifies soil types and their spatial distribution into four classes. Latosol and podzolic is in one single class respectively. Regosol, litosol, and organosol are collected in one other class. The last class consists of alluvial and gleysol.

Land use in this region varies. People use upland area as forests. Agriculture takes place in the form of mix garden in the middle region and rice fields at the lowland along the river. Settlements are distributed along the middle to lowland region. The survey on infiltration took the location in each soil class distribution area with two measurements at each point. It was found that the maximum level of infiltration was 2.098 mm for all soil types and land use, except settlements and wetlands. These two land uses give zero infiltration levels since they are impervious or saturated.

12.3.2 Government Response to Floods

The government has several agencies whose responsibilities are related to floods: the Geophysical and Meteorology Station; the station of river water level; the Water Resources Agency; public works; and the Social Welfare, Flood, and Natural Hazard Mitigation Office. The Geophysical and Meteorology Station provides records on weather data such as rainfall, temperature, evaporation, humidity, and wind velocity. The station of river water level records average the water level of the river. Some stations measure these parameters using manual tools and others use automatic recorders. The records are disseminated to the Water Resources Agency to be analyzed and to produce hydrologic data of this region. Some simple geographical information system (GIS) analysis such as delineation of watershed, digitizing of rain gauge station locations, contour mapping, and mapping of flood prone areas are done. Based on these data, areas that need structural development are determined.

Next, public works is responsible for structural development. In this regard Nikken Consultant, a foreign consultant from Japan, was hired to prepare a feasibility study and design for this structural development. Then the physical construction became the responsibility of the public works agency. It was completely finished in the year 2000 but has significantly reduced floods since the year 1998.

Finally, the Social Welfare, Flood, and Natural Hazard Mitigation Office was responsible for guaranteeing the safety of the community from any hazards, including floods. They began recording floods in 2003 and mapped the flood inundation area in 2005. They provide two boats to evacuate inhabitants from flooded areas. Since 2006, they have had one station located in the flood area to monitor flood elevation and prepare for evacuation.

12.3.3 Community Response to Floods

Data on community preparedness in facing floods were collected through in-depth interviews with community members in the flooded area. Inhabitants in the Air Dingin watershed are composed of indigenous people who are defined as people who have been living there for generations and newcomers who come from other regions and have lived here since the structural development on flood protection was put into operation in 1998. These indigenous people inherit knowledge to prepare for floods from their parents, but newcomers do not. Nonindigenous people who lived in this region before the structural development usually moved out to other areas after experiencing one period of monsoon floods. Therefore, there is a crucial difference between community preparedness for flooding before and after the structural development. Before the structural development, communities were aware and being prepared for floods. Floods usually come in the afternoon or early in the morning before school when children are at home. Almost nobody is outside their houses at these two times. Everybody now is alert.

People have various strategies to survive in flood events. Their houses are built with more than one floor or with higher ceilings with a door, used as a safe place of refuge during a flood. They usually track the rain visually, watching upstream if there is potential for a flood. Usually if a hard rain starts or even when there is no rain at all but sounds of swift flow are heard, they have 15 min to bring their valuables and survival things such as food, drugs, stoves, and blankets to the ceiling area before the flood. They will stay on the roof during the flood. They cook, eat, and sleep on the roof. Other activities cannot be done. If the water level keeps rising, personnel from the Social Welfare, Flood, and Natural Hazard Mitigation Office will come and bring two boats to evacuate them out of the area. To minimize losses, houses are built with elevated foundations. Each family prepares a table, sized $2 \times 2\,m^2$ with 1 m height, made of wood to save their furniture from a shallow flood. They also prefer to use wood furniture, which is less affected by flood water than is a fabric sofa. Their livestock are kept in elevated stalls. They learn from experience that the most important thing about livestock is their heads must stay above the water level. When the flooding ends, they clean their houses and all goods affected. Floods often carry stuff away and dumps waste into houses. Each family manages their own things. One family with neighbors coordinate in informing, warning, and sometimes in sharing the roof.

After the structural development, there was no preservation and enhancement of their previous awareness and preparedness in coping with a flood. Loss of local knowledge is accelerated by the arrival of newcomers who have no experience with a flood. On the other hand, economic development keeps rising and hydrologic conditions are becoming more difficult to predict.

12.3.4 The PCRaster Flood Forecasting Model

12.3.4.1 Synthesis of the PCRaster Flood Forecasting Model

Flood occurrence in the Air Dingin watershed was modeled based on the hydrologic characteristics of this region using PCRaster software, a prototype raster GIS for dynamic modeling that had been developed by Van Deursen (1995) as result of concepts behind the integration of dynamic models and GIS. Those inputs of the model were represented as layers and were analyzed to produce time series flood maps by executing scripts in a PCRaster window. PCRaster also has the capability to store values of a uniquely identified cell or cells written to a time series per time step into the ASCII format.

This capability makes it possible to calibrate and test the accuracy of a model based on the value of the observed water level. Some statistical parameters to evaluate the agreement between observations and forecasted flood shows that the mean absolute error (MAE) is −0.044, root mean square error (RMSE) is 0.144, correlation coefficient (CC) is 1.000, and coefficient of efficiency (CE) is 0.999. These parameters show that error of model is relatively small (±14%) and correlation between observed and forecasted water level is high (100%). It is expected that the calibrated model is relatively reliable to forecast dynamic flow of flood at Air Dingin.

Using this model, static maps of maximum flood prone areas for 10, 25, 50, 100, and 200 years of return periods were forecasted. Prediction of rainfall for those return periods were calculated by Gumbel modification, log Pearson type III, and the Iwai Kodoya method. The results of calculation are accepted if the Chi-square test shows that the degree of confidence from calculation is less than the theoretical one.

Using daily rainfall from the years 1979 to 2001, it was found that the maximum daily rainfalls accepted are achieved by the Iwai Kodoya method. It showed that maximum daily rainfall for 10 years is 324 mm/24 h, 25 years is 406 mm/24 h, 50 years is 474 mm/24 h, 100 years is 547 mm/24 h, and 200 years is 625 mm/24 h. Figs. 12.3–12.7 show these maps. These maps present a range of flood depth from zero to hundreds of meters.

12.3.4.2 Interpretation of the PCRaster Flood Forecasting Model

There are several reasons for error in the model. Flood models are developed using limited quantity and quality of hydrologic data on the Air Dingin watershed. The models use the best hydrologic data could be accessed. The results of models are calibrated using 30 daily records of water level at one station in the Air Dingin River.

The most reasonable reason for errors is the lack of a sampling point for calibration of water level data. But, as there is only one sampling point available for total calibration of water level in the entire catchment, accurate calibration for upstream, middle stream, and downstream water levels cannot be done. Moreover, it was found in the field that the quality of water level data is very low. The devices for measurement of the water level were not installed in the right way, water meters did not stand vertically as they should, but

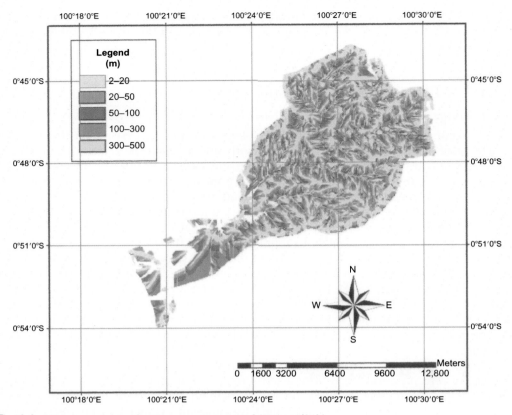

FIG. 12.3   Flood prone areas for 10 years return period (324 mm/24 h).

resembled, dragged along the river flow. No man stood in the station to control the devices. The responsible man only came to the station 4–5 times in a year. In accordance with this, field measurements on water level were only made 4–5 times/year. Daily records were interpolated from that limited data. Parametric uncertainty occurs as there are errors in the confidence intervals around the estimates from models as well as the measurement error in the data used to calibrate models. Improvement on data quantity and quality needs to be done if this data are to be used to produce a reliable flood forecasting model.

These errors need to be corrected before forecasts are calculated. As a result, the next spatial analysis needed in developing an warning system, including calculation of areas affected by floods, determination of affected population, affected land use, and so forth, cannot be done yet. To get valuable information for developing a flood warning system in each village, maps of village borders also have to be available.

Because loss of rain through evapotranspiration at the time of a rain event is very small and infiltration would stop after the saturation level is reached at 2.098 mm, it was decided that the main variables in developing a flood forecasting model would be rainfall record, water level record, and DEM.

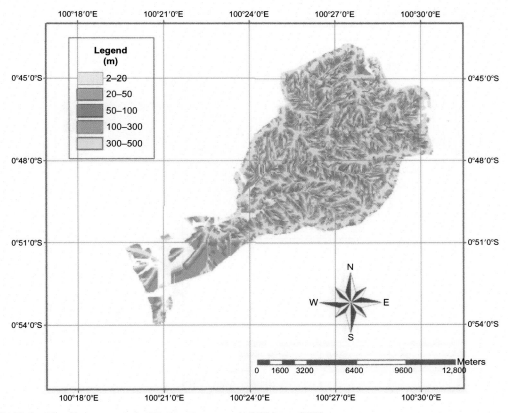

FIG. 12.4   Flood prone areas for 25 years return period (406 mm/24 h).

Real time rainfall intensity and duration are a crucial input to developing a flood forecasting model. It should be recorded on an hourly basis or even a minute-by-minute basis. Because there is usually too little time available from the start of rain to a flood occurrence, it would be so much better if rainfall forecasting is available. A preliminary model could be developed using this forecasting.

The matter of climatology development is becoming the responsibility of geophysical and meteorology stations. Analysis of the confidence interval between rainfall forecasting and rainfall field records should be done. Actually, detailed records on the intensity and duration of rainfall are available, but access to them is very limited.

Good quality of rainfall records cannot stand alone to build a reliable flood forecasting model. It should be synchronous with water level records in the Air Dingin River. The measurements should produce reliable data. Daily or hourly records should be available. DEM, as one of basic data sources for any kind of structural and nonstructural development, also needs to be improved and made more detailed.

On the other hand, the software package of the forecasting model also needs to be improved. Based on experience, this software package still has many limitations. It is in stages of continuous development by its developer, although many users have used it as part of

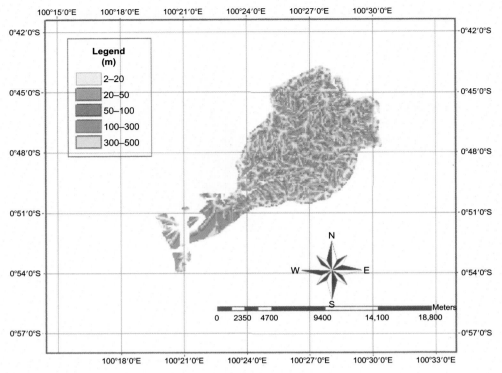

FIG. 12.5   Flood prone areas for 50 years return period (474 mm/24 h).

their decision support system. Some of the tools provided in the concept are not provided in the manual and cannot be executed. They include "traveltimestate" and "traveltimeflux," which present dynamic simulation in a more sophisticated way. However, the developer of this package provides a mailing list where they can inform and respond to users' questions. Some of other limitations of this software package are:

a. The type of data that could be stored in this software are limited into Boolean, nominal, ordinal, and scalar without possibilities to enter further description of properties for each cell in the raster maps.

b. The software does not recognize a south latitude coordinate system. It only recognizes positive values of coordinates. Tools for handling this case are provided, but unfortunately it does not function.

c. The dynamic model cannot be run in 3-D.

d. Any editing of properties in the legend display cannot be permanently saved for dynamic display. It has to be edited every time you need this information in your own preference classification.

The best way of evaluating the result is to verify the model with the real-world situation when the next flood hits the study area. However, the models are able to predict flood inundation areas, although the extent of floods are different.

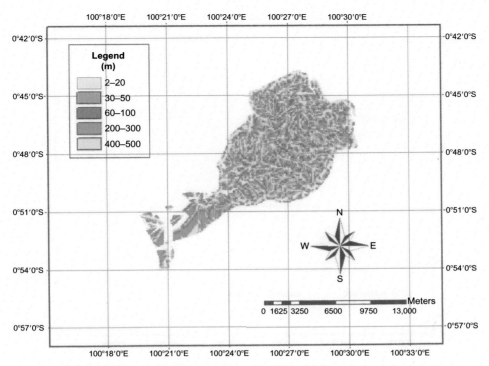

FIG. 12.6   Flood prone areas for 100 years return period (547 mm/24 h).

12.3.5 Synthesis of an Integrated Community-Based Flood Early Warning System

Because the Air Dingin watershed has more than 0.15 ha of arable area per capita, and scattered and low densities of population (<3.760 person/km^2) with low densities of economic development, there are still so many options of flood management strategy that could be taken. The strategies recommended include

a. Raising awareness.
b. Educating and enlarging access to information.
c. Developing community organizations.
d. Preserving local knowledge and strategy on flood preparedness.
e. Disseminating responsibilities to local representatives.
f. Training emergency actors.
g. Rehearsing emergency response.
h. Issuing warnings from hydrologic flood forecasting results.
i. Issuing warnings based on local knowledge.
j. Disseminating warnings to stakeholders,
k. Preparing for evacuation.
l. Providing reliable data for continuous improvement and hydrologic flood forecasting.

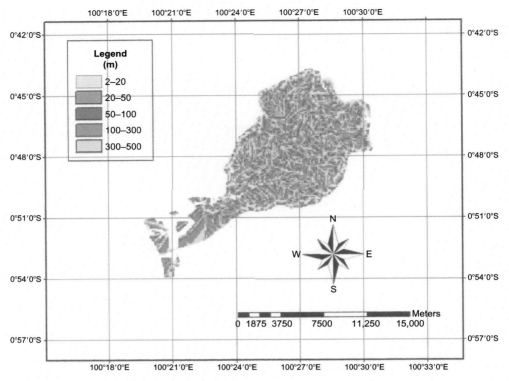

FIG. 12.7 Flood prone areas for 200 years return period (625 mm/24 h).

All these actions should involve governmental agencies, the media, the community, researchers, and nongovernmental organizations (NGOs). They include geophysics and meteorology stations, the Social Welfare, Flood, and Natural Hazard Mitigation Office, local authorities, police, medical and other emergency services, the media and water level stations, universities, voluntary organizations, an so on. Clear-cut roles and responsibilities, lines of command and control, and communication and coordination among these agencies need to be established. This can be achieved through regular meetings, joint training for emergency management staff, and rehearsals of emergency planning. Awareness and knowledge can be enhanced through public information programs provided by local agencies or NGOs. Publicity campaigns can involve public discussion, radio, and advertising, posters and leaflets, bulletins, newspapers, local television programs (Padang TV or TVRI Padang), and mobile campaigns by vehicles. A flood mitigation handbook need to be published, explaining about potential flood hazards, the government flood program, strategies before a flood, recommended flood proofing, action needed during a flood, action after a flood, website addresses, important phone numbers, and contacts.

It is required that any information that is made available to the public is in a simple form devoid of any technical jargon to ensure that the community understands and gets the messages. Access to information needs to be opened to as many organizations/institutions/persons as possible. There should be mechanisms for the public to give input, criticize, and contribute to the program. To achieve this objective the *Balai Pengelolaan Daerah Aliran Sungai* (Watershed Management Body) (BPDAS) needs to set up its own website or create an interactive program through radio and invite the public to participate to enhance continuous improvements on the existing strategies. It will yield transparency about what has been done by government, what agency is responsible for it, why the community needs to participate, and when and where they have to do it, how to do it, who funds it, and so forth.

The local community would know that their government cares about them. Support then will be given to the program. It will create bigger chances for the community to confirm, give input, and contribute to the development. Researchers and academics could get better access to data and improve their research on floods. The community needs to develop their organization called a management agency, where they can communicate, discuss, and set up their planning to cope with the next coming flood. They need to define their role in the program and enhance their capacity in coping with a flood. This management agency should be made up of representatives of those who live on or use the river rather than someone who is responsible to the central government and appointed by it. They are accountable downwards rather than upwards. Then, the government could transfer responsibilities related to floods to this catchments management agency; for example, in monitoring the function of existing structural development and issuing warnings if there is a possibility of a flood. Measurement of water level data could also be their responsibility. Gotong Royong (community cooperate and work together) to clean the river from dumped solid waste could also be done to keep capacities of the river in flowing floodwater. What had happened previously are patrols and inspections of any structural development by the responsible governmental agency cannot be done because of the lack of personnel and funds. The solution could be to include the cost of operations and maintenance in the proposed project fund and share responsibility with the community to do the job.

The Social Welfare, Flood, and Natural Hazard Mitigation Office as the emergency actor needs to enhance their capacities and resources. Training includes mobilization of emergency personnel and resources, search and rescue activities, the provision of emergency shelter and emergency feeding arrangements, and/or evacuation. Periodic checking on resources is crucial. There may be a tendency for agencies to double count resources available for use in emergencies or to be unaware of resources that could be mobilized. Therefore, careful drawing up of inventories of resources (personnel, transport, and supplies) should be a regular part of the preparedness phase.

At the lead time to a flood, warnings have to be delivered to the floodplain occupants affected: businesses, institutions, service providers, and residents. Any warning from hydrologic flood forecasting or the community should be sent to the management agency. This management agency decides to disseminate or stop the warning. If it will be disseminated, it needs to inform the masjid, radio, and other media, as well as the people and the agency responsible for emergency. Warning could be through telephone or short message service (sms) and siren or public announcement. Some objectives are to

a. Alert people at risk to the need to listen for and seek out further information and advice on the emergency;
b. Encourage organizations and individuals to alert others at risk to the danger.

c. Initiate flood proofing activities on individual properties.

d. Trigger compulsory or advisory evacuation out of the flood risk area.

e. Trigger removal to a safe shelter within the flood risk area, together with emergency supplies such as food, water, clothing, and essential medicines.

f. Stimulate property-saving activities such as moving livestock, business stock, and household effects to a safe place.

To some extent, flood management can be included as part of a generic all-hazards plan in terms of a place for evacuation. Emergency actors should inform people about location, process, and means of evacuation and in the same way to raise their awareness. Emergency planning should be grounded in an understanding of local people's perceptions and needs in an emergency situation. Local people are unlikely to wait passively for official advice but will act on their own initiative. Planning needs to take account of this condition.

Roles, responsibilities, and the proposed mechanism of flood forecasting and early warning are described in Fig. 12.8. This proposed mechanism relies on capacities of the local community and aims to preserve and enhance their knowledge with support from other governmental and nongovernmental parties. An effective and efficient mechanism is organized by minimizing stakeholders involved, because flood forecasting and early warning need real time hydrometeorologic data and needs to be disseminated and transformed into immediate action. The community has the biggest role in participation. However, there is

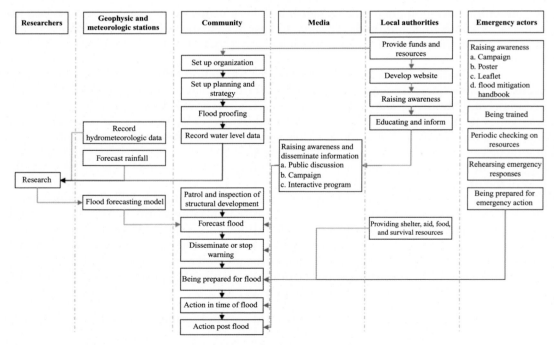

FIG. 12.8   Proposed scheme of roles and responsibilities of stakeholders in managing floods.

some elimination and transfer of roles and responsibilities of governmental agencies. To some extent, some tasks of the BPDAS, whose responsibilities are in developing better watershed management, including flood management, need to be transferred to meteorologic stations. After all, this scheme is not yet set in stone. Some improvement can be made to maximize achievement.

12.4 CONCLUSIONS

This research concludes that

1. The hydrologic condition of the Air Dingin watershed has the potential to monsoon flooding due to its characteristics:
   a. Time of concentration (time for storm flow to reach the outlet) is long. Huge and continuous rainfall needs to flow along a tiny and long watershed (watershed shape is like the feather of a bird extending from the upland to the estuary), creating high peak flow because the upper reaches of the watershed will be contributing runoff to the peak at the same time as the storm is over the outlet of the watershed.
   b. Curved river path blocks the flow, inundating adjacent area with lower elevation.
   c. High tide in monsoon months, blocking water flow into the ocean, creating water hammer to the inland.
2. Before structural development was put into operation in 1998, floods frequently came into the Air Dingin watershed in monsoon months. Extreme flow usually occurs in September, October, November, and December, and sometimes it continues to January. The highest flow usually comes in November when rainfall exceeds 100 mm/24 h. Flood levels are 1.5–2 m on average (medium-deep flood) and the areas is inundated for 2–5 days. After structural development completely finished in the year 2000, there was still 50 cm of flood on average (shallow flood) in the monsoon months.
3. Before structural development, the community in the Air Dingin watershed were aware and being prepared for floods. They develop adaptations to their behavior and the structure of their construction. Detailed explanation is provided in Chapter 5. After the structural development, there was no preservation and enhancement of their previous awareness and preparedness in coping with floods.
4. PCRaster, a GIS-based flood forecasting model was developed to make a scientific contribution to the early warning system. It produced dynamic flood models for the Air Dingin watershed with a 100% correlation level between forecasted and observed water level at the water level station on the Air Dingin River. But, unfortunately, it overestimated the real flood of the entire catchment by hundreds of meters. However, some considerations in the model were
   a. Data input to the models were topographic condition (DEM), precipitation (rainfall), and infiltration based on soil type and land use.
   b. Topographic conditions represented by DEM were compiled in the year 2000, derived from photogrammetric techniques (50–100 m spacing), consisting of contour lines. The map includes a geographic (latitude/longitude) reference, Universal Transverse Mercator, World Geodetic System 1984 datum Cent. Meridian 99d E.

   c. Rainfall was represented by records from two rain gauge stations: Tabing station (latitude 0°53′00″, longitude E 100°22′00″) and Kasang II station (latitude 0°53′28″, longitude 100°25′07″). Rainfall records were checked for quality by the double-mass curved technique using historical daily rainfall data from the years 1979 to 2001. The Thiessen polygon method shows that the Tabing station only represents 1.43% of total rainfall of this region and the Kasang II stations represent 98.57% of total rainfall of this region.

   d. Soil type in the Air Dingin watershed is constituted by latosol, podzolic, regosol, litosol, organosol, alluvial, and gleysol. Latosol and podzolic are classified as till soil. Regosol, litosol, and organosol are classified as outwash soil. Alluvial and gleysol are classified as wetland soil.

   e. Land uses in this watershed consist of forest, agriculture land, rice field, and settlements.

   f. Infiltration levels for different combinations of soil type and land use type are extremely the same. The maximum level is reached at 2.098 mm.

12.5 RECOMMENDATIONS

Based on the findings, the following recommendations are forwarded:

   i. Government needs to enhance and preserve community capacities in coping with floods.

  ii. Coordination and clear definition of roles and responsibilities between governmental agencies need to be defined immediately.

 iii. Reliability of hydrometeorologic data needs to be increased.

References

Abuzar, S.S., 2005. Potensi Sungai- Sungai Kota Padang sebagai Sumber Air Baku. Andalas University, Padang.

Aziz, F., 2001. A Dynamic GIS-Based Flood Warning System. Asian Institute of Technology, Thailand.

Statistical Bureau (BPS), 2005. Padang dalam Angka 2005. Statistical Bureau, Padang.

Statistical Bureau (BPS), 2006. Padang dalam Angka 2006. Statistical Bureau, Padang.

Van Deursen, W.P.A., 1995. Geographical Information Systems and Dynamic Models Development and Application of a Prototype Spatial Modelling Language. Faculty of Spatial Sciences, University of Utrecht, The Netherlands. http://www.pcraster.nl.

Rural Household Participation in Illegal Timber Felling in a Protected Area of West Sumatra, Indonesia

Yonariza*, E.L. Webb†

*Andalas University, Padang, Indonesia †National University of Singapore, Singapore, Singapore

13.1 INTRODUCTION

Timber harvesting from protected areas (PAs) is a threat to tropical forests, and is an international issue that has been attracting the attention of the international community because it is believed to cause environmental damage and promote corruption (Brack, 2005). Logging, particularly uncontrolled logging, can have variable but usually deleterious impacts on biodiversity and other globally important environmental services (Bawa and Seidler, 1998; Cannon et al., 1998; Putz et al., 2001; SCA and WRI, 2004). It may also contribute to increased poverty and social conflict (Tacconi et al., 2003).

In many PAs in Indonesia, timber harvesting activities are commonplace (Curran et al., 2004), especially since the collapse of a strong authoritarian government at the end of the 1990s (McCarthy, 2000; FWI and GFW, 2002; Hiller et al., 2004). The spread of illegal logging and other forest crimes into PAs occurs because valuable timber is still available in commercial volumes (Wardojo et al., 2001). Timber felling in PAs in Indonesia involve multiple stakeholders, including local people, logging companies, military personal, and forestry officials (McCarthy, 2002; Barber and Talbott, 2003; Laurance, 2004; Ravenel, 2004; Robertson and van Schaik, 2001; Hiller et al., 2004). Illegal logging is an opportunistic response to political change and bureaucratic confusion during transition in forest management policy as well as lack of appreciation of PA value (Newman et al., 1999, in Haeruman, 2001). Understanding the dynamics of illegal logging is an important step in finding sustainable solutions to these activities.

Local people involved in illegal logging are usually paid for their labor contribution and not as a function of the market value of the timber that they help extract (McCarthy, 2000). Nevertheless, illegal logging provides immediate income for local communities and may aid in day-to-day survival (Schroeder-Wildberg and Alexander, 2003). Moreover, logging activities such as chainsaw operating or timber hauling are highly risky and are generally considered an option of "last resort," when other livelihood opportunities are insufficient (Sunderlin et al., 2005; McCarthy, 2000). At best, it is seen as a route to quick comparatively substantial income.

In other parts of Indonesia, participation in illegal forestry activity is a function of local livelihood context as well as the need for the cash earned by this physically demanding, risky, and illegal activity. For example, a drop in income from subsistence crops led to increased illegal logging in Kutai National Park (East Kalimantan) and increased exploitation of non-timber forest products (NTFPs) in Lore Lindu National Park (Central Sulawesi) (Angelesen and Resosudarmo, 1999, in Murniati et al., 2001). The willingness of Indonesian villagers to engage in illegal logging was a function of several factors, including the need for income, whether other villagers (and nonvillagers) are already illegally logging, and the recognition of loss of community control over traditional forest areas (Dudley, 2004).

The Barisan I Nature Reserve, located in West Sumatra Province, is an important repository of biological diversity and a source of water for rural irrigation and nearby cities; yet, illegal logging takes place within its boundaries. Local people are the proximate vectors of timber felling in the nature reserve because they hold the chainsaws and haul the timber. Nevertheless, what are the forces that lead households around the nature reserve to engage in this activity? The considerable variety of factors presumed to underlie household decisions to participate in illegal logging, combined with a general paucity of systematic research on the subject, led to our site-specific research to improve understanding of the forces shaping local decisions about logging activities.

This research examined the practices of timber felling in the Barisan I Nature Reserve and had two objectives: (1) to describe the major timber-related activities undertaken by local households inside the nature reserve and (2) to analyze the contextual factors that lead to (a) participation in timber extraction activities and (b) the relative importance of timber-felling activities to the household income.

This study focused only on local (village and household) attributes and did not address externalities such as characteristics of the market that may put pressure on the households to participate. Our ongoing research has shown that not only are households the proximal agents through which all timber flows from the forest to the external markets, but also that the households as part of the *nagari* (larger village) have the capacity to halt the flow of timber from the forest (Yonariza, unpublished data). Therefore the households and villages are not only the proximal participants in tree harvesting, but also are the direct agents through which illegal logging can be reduced. Households respond to the available market through buyers. Therefore, understanding the household characteristics that may lead to participation in illegal timber harvesting will greatly improve the ability to focus intervention strategies to reduce the incentive to participate in that activity, thereby contributing to long-term forest conservation in Sumatra.

13.2 METHODS

13.2.1 Study Area

The Barisan I Nature Reserve covers an area of 74,000 ha in central West Sumatra province (100°22′38.34″ E–100°35′38.42″ E and 00°32′00.21″ S–00°57′50.34″ S) (see Fig. 13.1). It was declared a forest reserve during the Dutch colonial period through issuance of Government

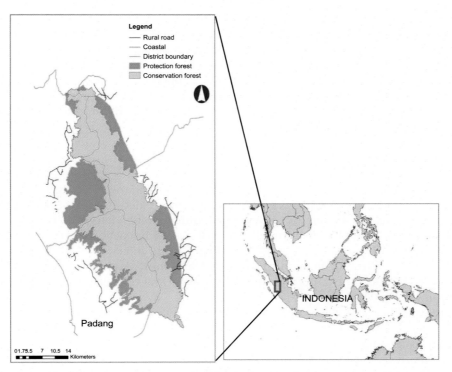

FIG. 13.1   Map of Barisan I Nature Reserve in West Sumatra Province of Indonesia showing study villages.

Blad No. 3 of Jan. 5, 1920. The central government established the nature reserve in consultation with the surrounding villages to ensure that the nature reserve covered only unoccupied land as per the Agrarian Law of 1870 (Jepson and Whittaker, 2002; Lindayanti, 2002), and that all stakeholders were aware of and agreed on the boundaries. This agreement was formalized in a letter between the council of elders and government officials. Thus, citizens of villages surrounding the nature reserve should be aware of its existence and extent.

At present, the nature reserve is classified as a category VI PA, which "should contain predominantly unmodified natural systems, managed to ensure long-term protection and maintenance of biological diversity, while providing at the same time a sustainable flow of natural products and services to meet community needs" (IUCN, 1994). The nature reserve plays a crucial role in ecosystem maintenance, particularly biodiversity conservation and water supply (Natural Resources Conservation Unit of West Sumatra Province [hereafter UKSDA Sumbar] 2000).

The nature reserve consists of conservation and protection zones (Fig. 13.1). In the conservation zone, no extractive activities can take place; in the protection zone, local people can harvest NTFPs. Timber cutting and forest conversion is prohibited in both zones. Local people refer to the nature reserve as *hutan larangan* ("prohibited forest") or *hutan lindung* ("protected forest"), phrases indicating that the nature reserve is protected from tree cutting and clearing. Villagers do not differentiate between the state-delineated conservation and protection zones in the nature reserve.

Elevation ranges from 100 m to approximately 2000 m asl (UKSDA Sumbar, 2000). Forest types are non-*Dipterocarpaceae* lowland, submontane, montane, and subalpine. Some commercially valuable timber is found in all forest types, so timber cutting can occur at any elevation.

The nature reserve is surrounded by densely populated villages in Solok, Padang Pariaman, Tanah Datar, and Padang City districts (UKSDA Sumbar, 2000). Eleven subdistricts and 23 *nagari*s (traditional villages) are immediately adjacent to the nature reserve where irrigated rice farming is the main livelihood activity, and residents directly or indirectly utilize nature reserve products for their livelihoods.

13.2.2 Data Collection

We purposively selected 11 *nagari*s in four districts (Padang, Padang Pariaman, Tanah Datar, and Solok), where villagers frequently participated in forest-related activities. Subsequently, a total of 17 *jorong*s (subvillages) immediately adjacent to the nature reserve were purposively selected, with approximately 10% of the households in each *jorong* ($n = 299$) randomly selected for survey (Pape, 1993; Gallego, 2005). Interview data for 60 households were destroyed in a fire, so 60 replacement interviews took place in a different *jorong*, as we felt uneasy to recollect the data from the same *jorong*. Data collection began in Jul. 2004 and lasted until May 2005.

Each selected household was visited and a face-to-face interview conducted. Primary data were collected through a questionnaire administered during household interviews. We were aware that discussing timber felling in the nature reserve was a sensitive topic, so discussion on timber was initiated only during the last part of the interview when a good rapport had been developed between respondent and interviewer. The questionnaire consisted of six parts, focusing on 18 key variables related to household livelihood context and timber-felling activities (Table 13.1).

Key informant interviews and forest observation data supplemented the household survey data and were used to describe timber-felling practices. Key informant interviews were conducted with officials from the *Unit Konservasi Sumber Daya Alam* (Natural Resources Conservation Unit) (UKSDA) of the Ministry of Forestry ($n = 4$), district level officials ($n = 6$), subdistrict (Kecamatan) administrators ($n = 10$), village administrators ($n = 25$), subvillage heads ($n = 17$), and forest farmer groups ($n = 2$). Forest observations were made in forests accessed by the villages to obtain physical evidence of timber felling.

13.2.3 Data Analysis

A household was categorized as a timber-felling household (TFHH) if it had a least one family member participating in cutting and sawing timber, hauling timber from the PA, or both. Timber trading was not included in the definition of a TFHH. If no family member engaged in felling activities, the household was categorized as a non-TFHH.

To test for the influence of household attributes on participation in timber-felling activities, we used logistic regression analysis (Mahapatraa and Kant, 2005) where involvement in timber felling was the binary dependent variable. We selected 16 contextual parameters as independent variables (Table 13.1) and performed a principle component analysis (PCA) with Kaiser normalization to reduce the number of independent variables to factors (Hair et al., 1998). PCA is powerful to simplify complex multidimensional variables into uncorrelated

TABLE 13.1   Research Variables on Factors Affecting Household Participation in Illegal Timber Felling in the Barisan I Nature Reserve, West Sumatra, Indonesia

Variables	Type	Description and Justification
Household size	Scale	The total number of household members present during the study. Larger households may have a larger labor force and therefore a higher probability of engaging in timber felling
Number of children being schooled	Scale	Although school fees are generally low, other schooling expenses are incurred and therefore may influence a household's need to engage in income-generating activities like timber felling or hauling
Number of outmigrant members	Scale	Family members emigrating out may contribute to the total household income through outside remuneration
Number of government subsidies received	Ordinal, 0=no subsidy received, 1=one subsidy received, 2=more than one subsidy received.	Subsidies were rice, health services, or scholarships. More subsidies indicate a lower income status and a higher likelihood to engage in timber felling
Number of high wealth objects	Scale	The total number of cars, trucks, vans, and motor bikes in the household. This is an indicator of wealth that was expected to be negatively associated with dependency on timber
Number of buffalo	Scale	Buffalo could affect involvement in timber felling in two ways. As buffalo is used to haul timber from forest, the possession of buffalo could indicate a household participates in timber felling. On the other hand, buffalo represent alternative income that could reduce dependency on timber
Involvement in forest farming	Dummy, 0=no, 1=yes	Forest farming is the establishment and use of temporary (eg, swiddening for annual crops) or permanent (planting of perennial crops such as fruit trees), small-scale agricultural plots inside the forest. Having a forest farming plot could indicate low output in permanent agriculture, thereby increasing the probability of a household engaging in timber felling
Involvement in firewood collection	Dummy, 0=no, 1=yes	Collection of firewood could indicate high dependence on forest products and a high engagement in other forest-related activities

Continued

TABLE 13.1   Research Variables on Factors Affecting Household Participation in Illegal Timber Felling in the Barisan I Nature Reserve, West Sumatra, Indonesia—cont'd

Variables	Type	Description and Justification
Involvement in hunting and trapping wildlife	Dummy, 0 = no, 1 = yes	Hunting wildlife for subsistence or market could be an activity associated with high dependence on forest products
Involvement in gathering nontimber forest products	Dummy, 0 = no, 1 = yes	Collecting nontimber forest products for subsistence or market could be associated with high dependence on forest products
Knowledge about existence of PA	Ordinal, 0 = the HH believes the reserve does not exist, 1 = the respondent is not aware of its existence, 2 = the HH is aware that the reserve exists	Scored the household according to its knowledge of the nature reserve. This is a combination of knowledge about the state-defined conservation and protection zones
Knowledge about legal status of logging in PA	Ordinal, 1 = respondent considers logging in PAs as legal, 2 = respondent does not know, 3 = respondent is aware that it is illegal	Knowledge about laws prohibiting logging within the nature reserve
Months of rice insufficiency	Scale	The months of rice production shortfall per year
Number of chainsaws	Scale	Presence and number of chainsaws may indicate the intensity or frequency of timber-felling activities
Presence of local forest use regulation	Dummy, 0 = no, 1 = yes	This variable measures whether the community has—and exercises—the authority to regulate forest use
Presence of local forest guard	Dummy, 0 = no, 1 = yes	Having a forest guard from the community and sponsored by the local government may decrease participation in illegal timber felling
Involvement in timber felling	Dummy, 0 = no, 1 = yes	Categorizes a household based on whether members are involved in cutting or hauling timber
Rank of income from timber	Ordinal, range 0–5	Respondent perceived importance of timber-felling activity in their household income, with five being most important

transformed factors (Cooley and Lohnes, 1971). This linear transformation has been widely used in data analysis and compression (Gonzalez and Woods, 1992).

We performed a backward stepwise logistic regression to test which contextual variables influenced the odds that a household would participate in timber felling. The logistic regression used the principle component scores as the independent variables, and the binary variable of involvement in timber felling as the dependent variable.

We conducted two linear multiple regressions to test the relationship between household characteristics and the importance of timber-felling activities to the household economy. We used the importance of timber income as the dependent variable because our respondent households did not practice bookkeeping, so it was not possible to acquire exact income quantities (Byron and Arnold, 1999). However, subjective income measures have been used with success in prior research (Das and van Soest, 1999; Jappelli and Pistaferri, 2000). The independent variables were the 16 contextual parameters plus the additional dummy variable of involvement in timber felling. We reduced the variables following the PCA procedure above, followed by a backward stepwise multiple regression. The component scores for each household were the independent variables, and the importance of timber felling was the dependent variable. For a second regression, we recalculated the PCA for the TFHH subset. This second regression evaluated which factors were related to a high importance of timber felling when households engaged in timber felling.

13.3 RESULTS

13.3.1 Timber Felling Practices

Approximately 19% of households were TFHHs and had a larger average household size, fewer outmigrants, and more schoolchildren than non-TFHHs (Table 13.2). The proportion of TFHHs receiving government subsidies such as rice, health services, and scholarships was higher than non-TFHHs. TFHHs had more months of rice insufficiency and a higher proportion of members involved in forest farming, hunting wildlife, and gathering NTFPs in the nature reserve, than TFHHs. Yet, both groups exhibited good knowledge on the existence of the nature reserve and the illegality of timber cutting within.

Since 1985, all cutting and sawing in reserve was done with chainsaws, replacing the hand-saw. Timber hauling was carried out manually (by 46 households), using water buffalo (four households), or using a tricycle (one household). In some cases, hauling required water transportation via river or canal. Timber was hauled from the forest to a point accessible by truck. From this collecting point, sawn timber was transported to shops and for sale to end users, while sawn logs were sent to a nearby sawmill. Hauling times varied from 0.5 to 7 h, depending on the tree species cut, with high-quality trees more distant and low timber quality trees near the forest edge.

Five TFHHs practiced cutting and sawing, 37 practiced hauling, and 14 were involved in both activities. Hence, 19 households were involved in cutting and sawing and 51 households were involved in hauling. Among the 19 cutting and sawing households, 5 participated on a regular basis (4 days per week, year-round), 3 seasonally (4 days per week, between rice planting and harvesting), and 11 only when there was an order for timber. Among 51 households involved in hauling, 22 households participated regularly, 8 seasonally, and 21 incidentally. Of the 56 TFHHs, 16 (29%) ranked timber first in household income (Online Supplementary Table 1).

Half of the TFHHs did not own a chainsaw, and the household member would be hired temporarily to cut timber. In this case the patron would supply the chainsaw, pay for its operation, and pay the timber cutter and hauler a labor rate based on the volume of timber. The labor rate ranged from 100,000 to 400,000 IDR/m^{-3}, depending on the tree species and distance of hauling. Because cutting and hauling 1 m^3 of timber could not be finished by one person in one working day, the per diem income would be about 15,000–40,000 IDR (approximately

TABLE 13.2   General Characteristics of Non-TFHH and TFHH Near the Barisan I Nature Reserve, West Sumatra

Household Characteristics	Non-TFH ($n=243$)	TFHH ($n=56$)
Average household size	5.3 (2.2)	6.0 (2.1)
Average number of children being schooled	1.34 (1.3)	1.7 (1.2)
Average number of outmigrant members	1.0 (1.8)	0.6 (1.2)
Recipient of rice subsidy	26.3%	32.1%
Recipient of health services subsidy	15.6%	28.6%
Recipient of scholarship subsidy	23.5%	28.6%
Average months of rice insufficiency	4.8 (4.7)	6.9 (4.6)
Mean of government subsidies received	0.7 (0.9)	0.9 (0.9)
Involvement in forest farming	63.0%	71.4%
Involvement in hunting and trapping wildlife	14.4%	26.8%
Involvement in gathering nontimber forest products	14.4%	25.0%
Knowledge about existence of PAs		
Does not exist	16.9%	16.1%
Exists	71.2%	69.6%
Do not know	11.9%	14.3%
Knowledge about legal status of logging in PA		
Not applicable (no knowledge about existence of PA)	15.2%	12.5%
Legal	5.3%	3.6%
Illegal	63.4%	60.7%
Do not know	16.0%	23.2%
Presence of local forest use regulation	16.5%	3.6%
Presence of local forest guard	7.8%	5.4%

Figures are mean (with standard deviation) for scale variables or percent of households for binary and ordinal variables.

US$1.50–4.00. The other half of TFHHs owned their own chainsaws and could either respond to an order, or cut trees independently and speculate on buyers. In this case the timber cutters and haulers would be paid according to the local market price of the timber. At the time of the study, the local market price for timber ranged from 800,000 IDR to 1,200,000 IDR/m^{-3}. Respondents expressed interest in owning their own chainsaws because of the potential income from freelance tree cutting.

13.3.2 Factors Affecting Household Participation in Timber Felling

Household attributes influenced household engagement in timber felling. PCA analysis on 16 independent variables returned 7 components that explained 64% of the total variance (Online Supplementary Table 2). These components were household size and number of government subsidies received, the presence of local forest control, knowledge about the existence of the nature reserve, absence of high wealth possessions and collection of NTFPs, involvement in forest farming, the number of buffalo, and the possession of a chainsaw and involvement in hunting wildlife. Results from logistic regression showed that household size and number of government subsidies received, absence of high wealth possession and involvement in gathering NTFPs, and possession of a chainsaw and involvement in hunting wildlife positively affected the odds of a household participating in timber felling (Table 13.3). Meanwhile, the presence of local forest control and the number of buffalo were negatively associated with the odds of a household participating in timber felling. Knowledge about the existence of the nature reserve and involvement in forest farming did not affect the odds of a household participating in timber felling.

Household attributes also influenced the income rank from timber felling. PCA on the 17 dependent variables returned seven components that explained 62% of total variability of the original data (Online Supplementary Table 3). These seven components included household size and number of subsidies received, the presence of local forest control, knowledge about the existence of the nature reserve, involvement in gathering NTFPs, involvement in timber felling and hunting wildlife, involvement in forest farming, and the number of buffalo. Regression analysis identified four factors that affected income rank from timber felling across TFHHs and non-TFHHs. Household size and number of subsidies received, and involvement in timber felling and hunting wildlife were positively associated with timber income rank, whereas involvement in NTFPs extraction and involvement in forest farming were negatively associated (Table 13.4).

Among TFHHs, only knowledge of the existence of the PA affected the income rank of timber. The PCA for 16 variables for the TFHH subsample returned seven components that explained more than 72% of the variance. These seven components included the presence

TABLE 13.3 Results of a Backward Logistic Regression Analysis Analyzing the Influence of Seven PCA Scores on Involvement in Timber Felling in the Barisan I Nature Reserve, West Sumatra

Factor	B	Std. Error	Wald	df	Sig.	Exp(B)
Constant	−1.802	0.194	86.474	1	0.000	0.165
Household size and number of government subsidies received	0.463	0.162	8.190	1	0.004	1.588
Presence of local forest control	−0.639	0.257	6.162	1	0.013	0.528
Absence of high wealth possessions and collection of nontimber forest products	0.368	0.180	4.172	1	0.041	1.445
Number of buffalo	−0.308	0.170	3.266	1	0.071	0.735
Possession of chainsaws and involvement in hunting and trapping wildlife	0.635	0.145	19.049	1	0.000	1.887

$n = 299$ households.

TABLE 13.4   Results of a Backward Stepwise Multiple Linear Regression using PCA Scores as the Independent Variables and Rank of Income as the Dependent Variable

Variable	Unstandardized Coefficients		Standardized Coefficients		
	B	Std. Error	Beta	t	Sig.
(Constant)	0.656	0.076		8.601	0.000
Household size and number of subsidies received	0.429	0.076	0.291	5.623	0.000
Involvement in NTFP extraction	−0.224	0.076	−0.152	−2.933	0.004
Involvement in timber felling and hunting	0.435	0.076	0.295	5.701	0.000
Involvement in forest farming	−0.189	0.076	−0.128	−2.479	0.014

$n = 299$ households surrounding the Barisan I Nature Reserve, West Sumatra.

of local forest control and NTFPs collection, household size and the absence of high wealth possessions, knowledge about the existence of PA, involvement in firewood collection and possession of chainsaw, the number of buffalo and involvement in forest farming, low income with involvement in hunting wildlife, and outmigration of family members. The regression revealed that only knowledge of the existence of the nature reserve affected the rank of income from timber felling. This component was negatively associated with the dependent variable, indicating that the more a household was aware about existence of the nature reserve, the lower the income rank of timber.

13.4 DISCUSSION

13.4.1 The Need for Cash and Illegal Timber Harvesting

There have been few systematic attempts to understand why rural households engage in high-risk illegal timber felling and how important this activity is to their livelihoods. Realizing the importance of household contextual variables as underlying facilitators of illegal logging, the UK Department for International Development (DFID) encouraged the Department of Forestry of Indonesia to sponsor micro-level studies to look at the role of illegal logging in local economies (Colchester, 2006). This study, although not part of that program, may contribute to fulfilling that need.

The demand for cash has prompted households to engage in timber felling, particularly in households with more members, more schoolchildren, and generally low income levels that necessitate government subsidies for health, rice, and scholarships. This result corroborates well with previous findings in Indonesia. Simulation models by Dudley (2004) have suggested that income-seeking and job-seeking were the driving forces behind communities engaging in illegal logging in Indonesia. Byron and Arnold (1999) argued that timber is seen as a resource to be tapped in times of extreme need. Our research provides quantitative evidence that households, given few income earning opportunities, may respond with undesirable or illegal natural resources extraction activities.

The positive relationship between the number of schoolchildren and participation in illegal timber harvesting highlights an interesting conservation and development conundrum. Education may be a long-term mechanism to improve income opportunities and therefore standard of living (Psacharopoulos and Patrinos, 1994; Lindenberg, 2002). According to mainstream thinking, this should lower dependence on forest extraction activities. But in the communities surrounding the Barisan I Nature Reserve, education is creating a short-term demand for cash to pay for childrens' school fees and daily expenses (eg, transportation). This demand is in some cases fulfilled by earnings from illegal logging activities.

Engagement in NTFP harvesting activities from the nature reserve is also a response to cash needs. However, the involvement in NTFP collection and other forest-related activities competes with time required for timber harvesting, explaining why the NTFP collection coefficient was positive for involvement in timber felling (Table 13.3), but negative in the rank income regression (Table 13.4). While it has been mentioned in the literature that labor is the main input in forest harvesting (Tropenbos International, 2005; Shackleton and Shackleton, 2003), the variable nature of NTFP collection's influence on logging via labor has not been considered.

Taken together, the results strongly suggest that the decision of a household to engage in timber harvesting activities is a response to a need for cash. Households also engage in other forest-related activities in an attempt to fulfill very basic livelihood requirements. It therefore follows that efforts to reduce income shortfalls could positively contribute to a reduction in illegal timber harvesting activities. Sunderlin et al. (2005; also Mulley and Unruh, 2004) suggest several factors that could generate income and thereby reduce pressure on the forest, such as increased agricultural productivity to reduce the profitability of agriculture in marginal areas, and greater off-farm employment to increase the opportunity cost of labor that might otherwise clear forests.

Results of the logistic regression analysis suggest that livestock raising could be a mechanism to reduce the participation of households in timber felling. Earlier studies have suggested the possibility of livestock to reduce dependence on forest resources in PAs (Sunderlin et al., 2005; Tropenbos International, 2005). This activity would not only generate income but also provide nutrition (Singh et al., 1985). However, investment in livestock could generate secondary impacts such as reduction of forest regeneration by forest grazing, fodder collection, or increased NTFP harvesting. Moreover, livestock such as buffalo do not earn daily cash. Other livestock activities such as poultry raising and cow milking, both of which provide daily income, could reduce the dependency on timber.

Intensification of agroforestry outside the nature reserve—expanding multipurpose home gardens on marginal and perhaps even cropland—could also reduce the need for timber felling. Murniati et al. (2001) found that in the Kerinci Seblat National Park of Sumatra, households with farms containing mixed perennial gardens were less dependent on park resources than households with only rice farming. Multiple products from agroforestry could provide sustained income by accessing multiple markets throughout the year. Moreover, agroforestry requires substantial labor and therefore would reduce the labor available for timber felling. In addition, Indonesian multistory agroforestry systems often have a complex structure and high biodiversity (Diemont and Martin, 2005; García-Fernández and Casado, 2005), maintain forest cover (Beukema and van Noordwijk, 2004), and increase habitat for wildlife (Nyhus and Tilson, 2004). Similar findings were reported in several African countries, where agroforestry enhanced and stabilized rural livelihoods, reduced pressure on PAs, enhanced habitat for some wildlife species, and increased the connectivity of landscape components (Ashley et al., 2006).

13.4.2 Local Control

Previous studies in PAs in Indonesia have found that timber networks include local people, timber traders, forestry officials, local police, military personnel, truckers, and local politicians (Schroeder-Wildberg and Carius, 2003; Smith et al., 2003; McCarthy, 2000). A popular recommendation has been to increase law enforcement at national and international levels (EIA and Telapak, 2006; Akella and Cannon, 2004). International entities such as the Consultative Group on Indonesia (CGI), a group of lending countries to Indonesia, encouraged the Indonesian government to enhance law enforcement to curtail illegal logging. The World Bank supported the Forest Law Enforcement and Governance (FLEG) process, and the European Commission launched the European Action Plan on Forest Law Enforcement, Governance, and Trade (FLEGT). Several attempts were made to enhance the law enforcement approach during each administration since 1999. Indonesia appears to follow the mainstream argument of increased enforcement for PA conservation.

Our fieldwork and interviews indicate that such policies are not entirely effective. When a state law enforcement officer detains a small-scale illegal timber harvester, it is difficult to sanction him according to the national policy. Article 78 point 4 of Law No. 41/1999 of the Forestry Law stipulates that the fine for illegal cutting is 5 billion IDR, more than US$500,000. This is equivalent to about 390 years of wage labor and therefore is never applied. Law enforcement ends up freeing those arrested, after which the activities continue as usual. Thus, state enforcement appears to be an ineffective method of conservation.

Our findings support enforcement if it is locally initiated and implemented. The presence of local forest control decreased the participation of households in timber-felling activities (see also McCarthy, 1999, for the case of Leuser National Park in Aceh Province). Local institutions may include community forest use regulations and the existence of community forest guards. Our findings agree with previous research showing that community support for sustainable forest management could reduce illegal logging (Dudley, 2001, 2004) or more generally, sustainable resource management (Ostrom, 1999; Gibson et al., 2005). In West Sumatra, local communities have traditional village systems (*nagari*) that have long histories of community-based forest management, so involving local people who have traditional rights over forest access to protect the nature reserve should lead to improved conservation.

An appropriate policy mechanism should be put in place for local people to participate in and have substantial management authority over PA conservation and management. Some authors have proposed comanagement for PAs (The World Bank, 1999; McCay and Hanna, 1998); our results support that strategy. Despite unresolved issues of comanagement, such as how to provide clear incentive mechanisms for all actors to get involved, comanagement appears to offer the greatest promise for long-term protection and enforcement of the Barisan I Nature Reserve.

13.4.3 Awareness: Important But Not Sufficient

Knowledge about the existence of the nature reserve did not affect the odds of a household engaging in timber harvesting, but did reduce the income rank in TFHHs. Together, these two results suggest that increasing awareness of the PA might contribute to the reduction of logging activities, but would not be an appropriate stand-alone method to prevent illegal logging. This is because the antilogging messages of educational programs cannot supersede the short-term need for cash, which can be fulfilled through timber harvesting activities that have a low risk of resulting in severe penalties from state enforcement agencies. Thus, while in general our results corroborate the International Union for Conservation of Nature (IUCN) recommendation to raise local community awareness and improve PA management (see Hamú et al., 2004), we argue that it will only achieve satisfactory results when part of a larger effort to improve livelihoods and localize control over forest resources.

13.5 CONCLUSION

When compared with Gunung Palung National Park, West Kalimantan, where 40% of the households had members who stated that logging was their primary source of cash income (Hiller et al., 2004), the problem of illegal logging in the Barisan I Nature Reserve appears to be less formidable. Our results were similar to other studies of rural households engaging in illegal logging in PAs: it is demand for cash income, usually during the off-farm season, which drives households to get involved. Our results suggest that households with fewer options for cash income are more likely to engage in illegal logging. Following Byron and Arnold (1999), these households are categorized as "those for whom forest dependency is a livelihood of last resort ... which they will abandon as soon as any plausible option emerges."

The key solution to the problem seems to be finding alternative sources of livelihood, particularly during off-farm work seasons. Many of our respondents in TFHHs claimed, "If we have other alternatives of work, we would stop timber felling." Development policy that seeks to provide livelihood alternatives to TFHHs could reduce local dependency on timber and at the same time contribute to conservation of the nature reserve. Additionally, forest conservation outcomes should improve if cooperation with local people in forest protection is developed, and if control over the resource is devolved to them.

Acknowledgments

This research was supported by Grant No. 1050-0726 from the Ford Foundation office in Jakarta to Andalas University. Assistance in the field was provided by Supomo, Novrianti, Yusmarni, P. Akhriadi, and F. Atmaja. An earlier version of this chapter was published in "Rural household participation in timber felling in a protected area of West Sumatra, Indonesia. 2007. *Journal Environmental Conservation* 34 (1): 73–82, Cambridge Univ. Press" which is duly acknowledged.

References

Akella, A.S., Cannon, J.B., 2004. Strengthening the Weakest Links Strategies for Improving the Enforcement of Environmental Laws Globally: CCG Report. Center for Conservation and Government at Conservation International, Washington, DC.

Ashley, R., Russell, D., Swallow, B., 2006. The policy terrain in PA landscapes: challenges for agroforestry in integrated landscape conservation. Biodivers. Conserv. 15, 663–689.

Barber, C.V., Talbott, K., 2003. The chainsaw and the gun: the role of the military in deforesting Indonesia. J. Sustain. For. 16, 137–166.

Bawa, K.S., Seidler, R., 1998. Natural forest management and conservation of biodiversity in tropical forests. Conserv. Biol. 12, 46–55.

Beukema, H., van Noordwijk, M., 2004. Terrestrial pteridophytes as indicators of a forest-like environment in rubber production systems in the lowlands of Jambi, Sumatra. Agric. Ecosyst. Environ. 104, 63–73.

Brack, D., 2006. Illegal Logging: Energy, Environment & Development Programme Briefing Paper EEDP/LOG BP 06/01 AUGUST 2006. London: Chatham House.

Byron, N., Arnold, M., 1999. What futures for the people of the tropical forests? World Dev. 27, 789–805.

Cannon, C.H., Peart, D.R., Leighton, M., 1998. Tree species diversity in commercially logged Bornean rainforest. Science 281, 1366–1368.

Colchester, M., 2006. Justice in the Forest, Rural Livelihoods and Forest Law Enforcement. Center for International Forestry Research (CIFOR), Bogor. 98 pp.

Cooley, W.W., Lohnes, P.R., 1971. Multivariate Data Analysis. John Wiley & Sons, New York. 364 pp.

Curran, L.M., Trigg, S.N., McDonald, A.K., Astiani, D., Hardiono, Y.M., Siregar, P., Caniago, I., Kasischke, E., 2004. Lowland forest loss in PAs of Indonesian Borneo. Science 303, 1000–1003.

Das, M., van Soest, A., 1999. A panel data model for subjective information on household income growth. J. Econ. Behav. Organ. 40, 409–426.

Diemont, S.A.W., Martin, J.F., 2005. Management impacts on the tropic diversity of nematode communities in an indigenous agroforestry system of Chiapas, Mexico. Pedobiologia 49, 325–334.

Dudley, R.G., 2001. Dynamics of illegal logging in Indonesia. In: Pierce, C.J., Resosudarmo, I.A.P. (Eds.), Which Way Forward? Forests, Policy and People in Indonesia. RFF Press, Washington, DC, pp. 358–382.

Dudley, R.G., 2004. A system dynamics examination on the willingness of villagers to engage in illegal logging. J. Sustain. For. 19, 31–53.

EIA, Telapak, 2006. Behind the Veneer, How Indonesia's Last Rainforests Are Being Felled for Flooring. EIA briefing, EIA and Telapak, London/Bogor. http://eia-global.org/images/uploads/Behind_The_Veneer.pdf.

FWI, GFW, 2002. The State of the Forest: Indonesia. Forest Watch Indonesia and Global Forest Watch, Bogor/Washington, DC.

Gallego, F.J., 2005. Stratified sampling of satellite images with a systematic grid of points. ISPRS J. Photogramm. Remote Sens. 59, 369–376.

García-Fernández, C., Casado, M.A., 2005. Forest recovery in managed agroforestry systems: the case of benzoin and rattan gardens in Indonesia. For. Ecol. Manag. 214, 158–169.

Gibson, C.C., Williams, J.T., Ostrom, E., 2005. Local enforcement and better forests. World Dev. 33, 273–284.

Gonzalez, R.C., Woods, R.E., 2008. Digital Image Processing, third ed. Pearson Prentice Hall Pearson Education, Inc. Upper Saddle River, New Jersey 07458. http://lit.fe.uni-lj.si/contents/students/publications/8/Gonzalez_DIP_predogled.pdf

Haeruman, H.J., 2001. Financing integrated sustainable forest and protected areas management in Indonesia: alternative mechanisms to finance participatory forest and protected areas management. In: International Workshop of Experts on Financing Sustainable Forest Management, Oslo, Norway, 22–25 January, 2001. A Government-Led Initiative in Support of the United Nations IPF/IFF/UNFF Processes.

Hair Jr., J.F., Anderson, R.E., Tatham, R.L., Black, W.C., 1998. Multivariate Data Analysis, fifth ed. Prentice Hall, Upper Saddle River, NJ.

Hamú, D., Auchincloss, E., Goldstein, W. (Eds.), 2004. Communicating Protected Areas, Commission on Education and Communication. IUCN, Gland/Cambridge. 312 pp.

Hiller, M.A., Jarvis, B.C., Lisa, H., Paulson, L.J., Pollard, E.H.B., Stanley, S.A., 2004. Recent trends in illegal logging and brief discussion of their cause: a case study from Gunung Palung National Park, Indonesia. J. Sustain. For. 19, 181–212.

IUCN, 1994. Guidelines for Protected Area Management Categories. CNPPA With the Assistance of WCMC. IUCN, Gland/Cambridge. 261 pp.

Jappelli, T., Pistaferri, L., 2000. Using subjective income expectations to test for excess sensitivity of consumption to predicted income growth. Eur. Econ. Rev. 44 (2), 337–358.

Jepson, P., Whittaker, R.J., 2002. Histories of protected areas: internationalisation of conservationist values and their adoption in the Netherlands Indies (Indonesia). Environ. Hist. 8, 129–172.

Laurance, W.F., 2004. The perils of payoff: corruption as a threat to global biodiversity. Trends Ecol. Evol. 19, 399–401.

Lindayanti, R., 2002. Shaping local forest tenure in national politics. In: Dolsak, N., Ostrom, E. (Eds.), The Comons in the New Millennium. The MIT Press, Cambridge, pp. 221–264.

Lindenberg, M., 2002. Measuring household livelihood security at the family and community level in the developing world. World Dev. 30, 301–318.

Mahapatraa, K., Kant, S., 2005. Tropical deforestation: a multinomial logistic model and some country-specific policy prescriptions. Forest Policy Econ. 7, 1–24.

McCarthy, J.F., 1999. Village and state regimes on Sumatra's forest frontier: a case from the Leuser Ecosystem, South Aceh: Resource Management in Asia-Pacific Working Paper 26. In: Paper Presented in the Resource Management in Asia-Pacific Project Seminar Series, November 1999.

McCarthy, J.F., 2000. "Wild Logging": The Rise and Fall of Logging Networks and Biodiversity Conservation Projects on Sumatra's Rainforest Frontier: CIFOR Occasional Paper 31. Center for International Forestry Research, Bogor. Available from: http://www.cifor.cgiar.org/publications/pdf_files/OccPapers/OP-31.pdf (accessed 29.03.04).

McCarthy, J.F., 2002. Power and interest on Sumatra's rainforest frontier: clientelist coalitions, illegal logging and conservation in the Alas Valley. J. Southeast Asian Stud. 33, 77–106.

McCay, B.J., Hanna, S., 1998. Co-managing the commons: creating effective linkages among stakeholders: lessons from small-scale fisheries. In: Report from the International CBNRM Workshop, 10–14 May, 1998. The World Bank, Washington, DC. http://www.worldbank.org/wbi/conatrem/.

Mulley, B.G., Unruh, J.D., 2004. The role of off-farm employment in tropical forest conservation: labor, migration, and smallholder attitudes toward land in western Uganda. J. Environ. Manag. 71, 193–205.

Murniati, D.P., Garrity, A., Gintings, N., 2001. The contribution of agroforestry systems to reducing farmers' dependence on the resources of adjacent national parks: a case study from Sumatra, Indonesia. Agrofor. Syst. 52, 171–184.

Nyhus, P., Tilson, R., 2004. Agroforestry, elephants, and tigers: balancing conservation theory and practice in human-dominated landscapes of Southeast Asia. Agric. Ecosyst. Environ. 104, 87–97.

Ostrom, E., 1999. Self-Governance and Forest Resources: CIFOR Occasional Paper 20. Center for International Forestry Research (CIFOR), Bogor. 15 pp.

Pape, E.S., 1993. The cost of randomization in work sampling: an illustration. IIE Trans. 25 (6), 89–96.

Psacharopoulos, G., Patrinos, H.A., 1994. Indigenous People and Poverty in Latin America: An Empirical Analysis. Human Resources Development and Operations Policy, The Word Bank, Washington, DC. 20 pp.

Putz, F.E., Blate, G.M., Redford, K.H., Fimbel, R., Robinson, J., 2001. Tropical forest management and conservation of biodiversity: an overview. Conserv. Biol. 15, 7–20.

Ravenel, R.M., 2004. Community-based logging and defacto decentralization: illegal logging in the Gunung Palung area of West Kalimantan, Indonesia. J. Sustain. For. 19, 213–237.

Robertson, J.M.Y., van Schaik, C.P., 2001. Causal factors underlying the dramatic decline of the Sumatran orang-utan. Oryx 35, 26–38.

SCA, WRI, 2004. "Illegal" Logging and Global Wood Markets: The Competitive Impacts on the U.S. Wood Products Industry. Seneca Creek Associates, Wood Resources International, LLC, Poolesville, MD. 163 pp.

Schroeder-Wildberg, E., Carius, A., 2003. Illegal Logging, Conflict and the Business Sector in Indonesia. InWEnt-Capacity Building International, Berlin, Germany. 76 pp.

Shackleton, C., Shackleton, S., 2003. Value of non-timber forest products and rural safety nets in South Africa. In: Paper Presented at The International Conference on Rural Livelihoods, Forests and Biodiversity, Bonn, Germany, 19–23 May, 2003.

Singh, A.K., Singh, M.K., Mascarenhas, O.A.J., 1985. Community forestry for revitalizing rural ecosystems: a case study. For. Ecol. Manag. 10, 209–232.

Smith, J., Obidzinski, K., Subarudi, Suramenggala, I., 2003. Illegal logging, collusive corruption and fragmented governments in Kalimantan, Indonesia. Int. For. Rev. 5, 293–302.

Sunderlin, W.D., Angelsen, A., Belcher, B., Burgers, P., Nasi, R., Santoso, L., Wunder, S., 2005. Livelihoods, forests, and conservation in developing countries: an overview. World Dev. 33, 1383–1402.

Tacconi, L., Boscolo, M., Brack, D., 2003. National and International Policies to Control Illegal Forest Activities: A Report for the Ministry Of Foreign Affairs, Government of Japan. Center for International Forestry Research, Bogor. 63 pp.

The World Bank, 1999. Report from the International CBNRM Workshop, Washington, DC, 10–14 May, 1998. http://info.worldbank.org/etools/docs/library/97605/conatrem/conatrem/documents/May98Workshop_Report.pdf.

Tropenbos International, 2005. Alternative livelihoods and sustainable resource management. In: Inkoom, D.K.B., Kissiedu, K.O., Wageningen, Jr, B.O. (Eds.), Proceedings of a Workshop Held in Akyawkrom, Ghana, on the 1st of April 2005. Tropenbos International, The Netherlands, p. 71.

III. SOCIOECOLOGICAL SYSTEMS AND NEW FORMS OF GOVERNANCE

UKSDA Sumatra Barat, 2000. Rencana pengelolaan cagar alam Barisan I Propinsi Sumatra Barat (Management Plan Barisan I Nature Reserve in West Sumatra). Unit of Conservation and Natural Resources Protection West Sumatra (UKSDA) Sumatra Barat, Padang. 115 pp.

Wardojo, W., Suhariyanto, Purnama, B.M., 2001. Law enforcement and forest protection in Indonesia: a retrospect and prospect. In: Paper Presented on the East Asia Ministerial Conference on Forest Law Enforcement and Governance (FLEG), Bali, Indonesia, 11–13 September, 2001.

Yonariza is professor in forest resources management at Andalas University in Padang, West Sumatra, Indonesia. He obtained his PhD in Natural Resources Management from Asian Institute of Technology (AIT), Bangkok, Thailand, in 2007. He was an Asian Public Intellectuals Fellow from 2008 to 2009, supported by The Nippon Foundation, and carried out a research project on postlogging ban timber tree planting in Thailand. He is currently the chair of Integrated Natural Resources Management Field of Study, Graduate Program, Andalas University.

Edward Webb has lived in Southeast Asia since 1998. He was on the faculty of Natural Resources Management at the Asian Institution of Technology (Thailand) and is currently an Associate Professor in the Department of Biological Sciences at the National University of Singapore. His research interests cross multiple disciplines with the explicit goal of furthering our understanding of tropical forest ecology, management, and conservation. He specializes in vegetation ecology, tropical tree dynamics, community-based forest management, and conservation policy.

Decentralization and Comanagement of Protected Areas in Indonesia

Yonariza*, G. Shivakoti[†,‡]

*Andalas University, Padang, Indonesia [†]The University of Tokyo, Tokyo, Japan
[‡]Asian Institute of Technology, Bangkok, Thailand

14.1 INTRODUCTION

With the changing development paradigm in the recent past, the role of communities and local government have been reemphasized and the roles of central government have steadily decreased, changes that have been attributed to decentralization policies. Various options for natural resources management are being debated, including cooperative management (hereafter comanagement) among key stakeholders (the central government, local government, and local communities). There are several reasons for adopting comanagement: the failure of market approaches in resources allocation; the failure of government monopoly, or top-down decision making in resources management; and also the failure of community-based resources management (bottom-up decision making) due to lack of coordination with the state and the market.

Protected areas (PAs) management in developing countries has faced several issues, including the recent decentralization trend that presents a new challenge on how to make decentralization work for conservation processes such as PAs management (Lutz and Caldecott, 1996; Wyckoff-Baird et al., 2000; Andersson and Clark, 2004). In the past, local and regional governments have viewed PAs within their districts as obstacles to local government revenue generation. Thus they did not want to participate in conservation efforts because they were costly and they had no incentive to get involved in them (see Griffiths et al., 2002). In addition, these local and regional governments often do not share national and international concerns for biodiversity conservation (McCarthy, 2000). A clear challenge is to make local government a partner in PAs management.

Along with the paradigm shift in the relationship between PAs and local people, issues concerning the livelihood of local people have emerged, most substantially in conflicts on issues of partnership and collaboration and the management of context specific partnerships affecting all stakeholders (Rao, 2001; Elliott, 1996). Identifying appropriate institutions at local levels is another challenge. As da Silva (2004) argues, central to this process is the recognition and legitimization of traditional or informal local-level management systems.

Recent decentralization in Indonesia is a case in point. Studies of PAs management in Indonesia after the implementation of government decentralization policies have produced very controversial results. It is widely reported that decentralization triggers local government efforts to exploit the remaining forest resources, regardless of their status, with the object of earning short-term revenues either through timber cutting or by converting forest areas into agricultural plantations, or by converting protected forest into production forest to increase regional income from logging permits (Aden, 2001; Sudana, 2004; Dewi et al., 2005; FWI/GFW, 2002; McCarthy, 2004; Obidzinski, 2004; Obidzinski and Barr, 2003; Obidzinski, 2004; Rhee, 2000; Casson and Obidzinski, 2002). Illegal logging in PAs, involving many stakeholders including local people, logging companies, military personal, and corrupt forestry officials, is found to be increasing (McCarty, 2002; Barber and Talbot, 2003; Laurance, 2004; Ravenel, 2004; Hiller et al., 2004).

In Indonesia the concept of comanagement involving central government, local government, the commercial private sector, local communities, and civil societies (World Bank, 1999), in combination with the currently ongoing decentralization reform, has created a better environment for implementing this model as a means of effectively managing community resources. It remains to be seen, however, to what extent decentralization will trigger the creation of comanagement. Therefore, several issues arise, such as why decentralization and policy reform does not automatically produce comanagement, and what should be done to expand these reforms on a wide scale to include PAs comanagement. This chapter seeks to explore the important issues related to PA comanagement and the provision of structures of incentives for Indonesia. Specifically, in this chapter we examine whether locally available sources of finance and incentives for providing financing for PAs as matters of local interest have been sufficiently developed. Debate on the need to involve local communities in PAs management for the conservation of the remaining forest has been energized as these local communities are the most affected stakeholders and so have a keen interest in the enforcement of the regulation (Gibson et al., 2005).

Scholars have defined comanagement as "the sharing of power and responsibility between the government and local resource users" (Carlson and Berkes, 2005, citing Berkes et al., 1991: 12). Comanagement can succeed only when the incentives for local government and local communities to participate in PAs management are spelled out clearly. In addition, its success depends on the extent to which local people depend on forest resources and how far appropriate local institutions are partners with local communities

We examined these issues of comanagement through the study of the Barisan I Nature Reserve, a PA in West Sumatra Province, Indonesia. The study was conducted between August 2004 and May 2005, three years after the implementation of a government decentralization policy in the country. Timber felling has been rampant in this PA. The PA straddles four autonomous districts. Hence, the response of each district government toward the protection of this nature reserve under current decentralization is examined and the effects of the responses are compared. Before we present discussion on our research site we review the current decentralization and forest management policies in Indonesia to determine whether the current administrative system in the country is suitable for its propose, and we develop a comanagement model.

14.2 POLICY CHANGES AND PROSPECTS FOR PROTECTED AREAS COMANAGEMENT IN INDONESIA

Along with the global trend to decentralization, the current regional autonomy policy in Indonesia opens a possibility for comanagement in PAs involving the central government, local government, local communities, and civil society (see, for example, FWI/GFW, 2002; Haeruman, 2001; Clifton, 2003). Optimism about this has been reinforced by radical changes in forest policy, the reorganization of government administration through decentralization, and biodiversity conservation policy.

Forest governance and management in Indonesia, which is governed by Law No. 41/1999, explicitly includes provisions for decentralization and local people empowerment. In addition, the drafting of the law has been a participatory process through the involvement of civil society groups (Lindayanti, 2002). Thus it lays the foundation for a participatory approach to forest management in all areas including PAs.

Article 6.1 of the law outlines three main forest functions; namely, conservation, protection, and production. Production forest has the main function of producing forest products; protection forest on the other hand has the main function of protecting life-supporting systems for hydrology, preventing floods, controlling erosion, preventing sea water intrusion, and maintaining soil fertility; and conservation forests enclose an area with specific characteristics having the main function of preserving plant and animal diversity and its ecosystem. Protection forest and conservation forest fall under the category of PA. Further, conservation forest as elaborated in Article 7 is divided into (a) nature reserve forest, which means a forest with specific characteristics, having the main function of preserving plant and animal diversity and its ecosystem, and also serving as the place for life-supporting systems; and (b) nature preservation forest area, which means a forest with specific characteristics, having the main function of protecting life-supporting systems, preserving species diversity of plants and animals, and enabling the sustainable use of biological resources and its ecosystem. Our study sites belong to nature reserve forest.

The central government, however, still holds strong control over the forest under Chapter 1 of the law, especially in Article 4, which gives the central government the power to (i) regulate and organize all aspects of forest, forest areas, and forest products; (ii) determine the status of an area as a forest area or a nonforest area; and (iii) regulate and determine legal relations between man and forest, and regulate legal actions concerning forestry.

The role of local government in managing forest has been specified. In Chapter 8 of the law on the delegation of authorities, Article 66 specifically has stipulated that in implementing forest administration, the central government shall delegate part of its powers to local government. The roles of local government in this respect shall, however, be regulated by a central government regulation. In the implementation of this article, the government has issued three government regulations (hereafter GRs): GR No. 34/2002 on forest management and forest management planning, forest use, and utilization of forest area; GR No. 44/2004 on forest planning; and GR No. 45/2004 on forest protection. These regulations stipulate the role of local government in many aspects of forest management. Basically there is a delegation of authority to local government to manage forest areas within their jurisdiction by following guidelines provided by the central government.

The current forestry law also recognizes the role of local people in forest and PA management, thus providing room for local communities to participate. Chapter 9 of the law is devoted to community customary law. Similarly, Article 69 provides (a) that communities shall be obliged to participate in maintaining and preventing forest areas from disturbance and damage, and (b) that in implementing forest rehabilitation, communities can also request assistance, guidance, and support from nongovernmental organizations, other parties, or the government. Community roles are further elaborated in Article 70 emphasizing their importance in the comanagement of forestry resources.

A similar type of arrangement for the sharing of rights and responsibility in PA management among the central government, local government, and the local community can be found in PAs policy stipulated in Law No. 5/1990 regarding biodiversity and its ecosystem protection.

In the context of PAs management, it is worth mentioning that the decentralization policy enacted in 1999 and revised in 2004 has made provision for the central government to transfer to autonomous local governments authority to manage local resources including PAs. These have become an "optional obligation" for local government as not all regions have forest and PAs, and they can also be regarded as a compulsory obligation to control the environment.

However, as this chapter will show later, decentralization does not automatically bring about comanagement models. With regard to this, several issues need to be addressed, including the identification of local partners and the requirement of conditions conducive to the environment. We have examined these issues in a PA of West Sumatra Province, where different levels of decentralization and comanagement models are being implemented.

14.3 STUDY AREA

The Barisan I Nature Reserve in West Sumatra Province covers an area of 74,000 ha (Fig. 14.1). According to the International Union for Conservation of Nature (IUCN) PAs classification, this reserve area belongs to category VI. Areas in this category should contain predominantly unmodified natural systems managed to ensure long-term protection and maintenance of biological diversity while providing at the same time a sustainable flow of natural products and services to meet community needs. The IUCN has outlined the objectives of management PA category VI as (a) to protect and maintain the biological diversity and other natural values of the area in the long term; (b) to promote sound management practices for sustainable production purposes; (c) to protect the natural resource base from being alienated for other land-use purposes that would be detrimental to the area's biological diversity; and (d) to contribute to regional and national development (IUCN, 1994). The government of Indonesia since 1982, however, has considered this area as a nature reserve where no harvesting of forest products is allowed. It is intended to be the research equivalent of ICUN category I, a strict nature reserve. Since 2002, the Department of Forestry has even proposed that the Barisan I Nature Reserve be classed as a national park considering the importance of biodiversity and ecological functions in the area.

The reserve straddles four autonomous districts; namely, Padang City (the capital of West Sumatra Province), and Padang Pariaman, Tanah Datar, and Solok districts. These autonomous districts have implemented various levels of decentralization. Hence, the response of each district government to the protection of the Barisan I Nature Reserve in their respective jurisdictions under current decentralization can be examined and compared.

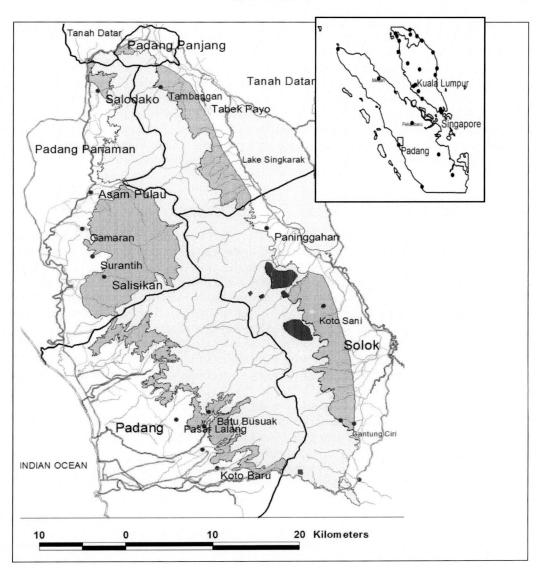

FIG. 14.1   Barisan I Nature Reserve, West Sumatra, Indonesia.

The Barisan I Nature Reserve has the environmental function of maintaining water conditions in the catchments of several rivers that supply water to Singkarak Lake where a 154 MG hydroelectric power plant operates. These rivers also supply water to a number of small-scale irrigation systems surrounding the forest reserve. Thus it has an important role in hydrological regulation. This is a long-established PA that dates back to the Dutch colonial time in the early 20th century. Thus, local understanding of its existence can be expected to be high.

The Barisan I Nature Reserve represents a continuous forest, which under the recent Forestry Law has a complex system of forest management. Its inner part is conservation forest, which by law is the responsibility of the central government but is surrounded by protection forest, which is under the authority of the district government. Further, communal forests are either under the management authority of the community (*nagari*, *suku*), or under extended family authority. We examine the important role played by each group of stakeholders in PAs management. The issue of management gets further complicated because this reserve area is surrounded by 23 *nagari*s (traditional villages) that have traditional claims of land rights inside PAs and whose roles in forest protection have been enhanced under the current government decentralization policy in West Sumatra Province.

14.4 SITE AND HOUSEHOLD SELECTION CRITERIA

Decentralization in West Sumatra Province has been implemented by the *return to nagari* policy. However not all districts have adopted it. We have divided the study area into four zones according to model and degree of decentralization. In cities especially, the policy is not implemented, and city government still follows the national administration model called *Kelurahan* as the lowest administrative unit. For our study purpose, we call this district and its type of decentralization category 1 (D-I). In other districts, there are varying degrees of decentralization implemented, with the revitalization of a *nagari* administration system and comanagement. There are districts with revitalized *nagari*s but without power decentralized to *nagari* government (D-II), districts with revitalization and decentralized power but not performing comanagement (D-III), and districts with revitalization and re-decentralization and also performing comanagement (D-IV).

Based on the above model and level of decentralization, and taking a number of prevalent forest-related activities (such as farming, fuel wood collection, nontimber forest products (NTFPs) collection, hunting and trapping, and persistent timber felling) in each *nagari*, we selected 11 out of 23 *nagari*s surrounding reserves for detailed study including household surveys. Three villages (*kelurahan*) belonged to D-I in Padang, another three *nagari*s belonged to D-II in Padang Pariaman District, two *nagari*s belonged to D-III in Tanah Datar District, and three *nagari*s belonged to the D-IV category in Solok District. From each selected *nagari*, *jorong*s (subvillages) having a border with the Barisan I Nature Reserve were purposively selected as study sites; the total number of these selected subvillages being 17. In each selected subvillage, approximately 10% of households were randomly selected for interview, giving a total of 299 households. Basic information about these households is presented in Table 14.1.

14.5 FINDINGS

14.5.1 Multilevel Decentralization

As mentioned earlier, Law No. 22, enacted by the government of Indonesia in 1999, served as a basis for regional autonomy. This law was intended to reduce the control of the centralized and authoritarian government through decentralization during the New Order Regime and to

TABLE 14.1 Some Basic Information About Respondents and Their Households

Level of Decentralization and Comanagement	Number of Sample	Respondents' Level of Education				# Respondents Involved in Farming	Average Household Size	Average Irrigated Land Holding (ha)	Average Upland Holding (ha)
		Illiterate	Elementary School	High School +					
D-I (no revitalization)	70 (100.0)	8 (11.4)	33 (47.1)	29 (41.4)		30 (42.9)	6	0.26	0.59
D-II (revitalization but no decentralization	75 (100.0)	2 (2.7)	57 (76.0)	16 (21.3)		45 (60.0)	5	0.41	1.11
D-III (revitalization, decentralization, but no collaboration)	74 (100.0)	5 (6.8)	37 (50.0)	32 (43.2)		46 (62.2)	5	0.64	0.86
D-IV (revitalization, decentralization, and collaboration)	80 (100.0)	6 (7.5)	52 (65.0)	22 (27.5)		58 (72.5)	6	0.45	0.88
Total	299	21	179	99		179			
	(100.0)	(7.0)	(59.9)	(33.1)		(59.9)			

Note: Figures in parenthesis indicate percentages.

III. SOCIOECOLOGICAL SYSTEMS AND NEW FORMS OF GOVERNANCE

acknowledge the social, political, and cultural diversity in the country. In West Sumatra, the process of decentralization has been of a particularly dynamic and interesting character, where along with the general decentralization of central political authority and economic resources to the districts, a fundamental restructuring of local village government has also been initiated (Benda-Beckmann and Benda-Beckmann, 2001). The policy of regional autonomy has been taken up with the aim "to return to the *nagari*." Some district governments, in turn, shifted some of their authority to *nagari* government. There is a great expectation that by implementing multilevel decentralization, government at all levels will be more responsive toward local needs and hence participation will increase (Benda-Beckmann and Benda-Beckmann, 2001).

However, local responses to decentralization vary widely both across spatial and infrastructural dimensions as well as at the macro, meso, and micro level. District heads and parliamentarians have (re)acted with different speeds and degrees of enthusiasm to these developments. Earlier research conducted in West Sumatra reported that in 2 out of 14 districts in the province (Limapuluh Kota and Solok), district heads had taken a number of initiatives to implement the new structure quickly by revising their district administrative structure and pushing forward the return to the *nagari* system. They promulgated their own district regulations, and Solok District implemented local administration changes integrating the *nagari* system as soon as the provincial regulation became effective in January 2001 (Benda-Beckmann and Benda-Beckmann, 2001). With regard to PAs management, Solok District has also moved much ahead of other districts by taking several initiatives that we discuss in detail in the following section.

14.5.2 District Government Initiative

As mentioned earlier, varying degrees of decentralization have affected local initiatives differently with respect to PAs comanagement. In D-I, where revitalization of the *nagari* has not been initiated, nothing has changed since the implementation of government decentralization. In D-II, revitalization of the local administration has not been followed up by decentralizing power, while in D-III revitalization of village administration has been followed by decentralization of power to village administration, but there has been no initiative for comanagement. It is only in D-IV that *nagari* revitalization has been followed by decentralization and formation of comanagement of PAs by district government and *nagari* government.

Solok District (D-IV) encloses several PAs, and the stakeholders have made a systematic effort to maintain these PAs. The District Forestry Service issued a decree on the establishment of Community Forest Guarding Units (CFGUs) in 2003. Along the Barisan I Nature Reserve, the district government set up CFGUs in four *nagari*s consisting of the head of the *nagari*, chief of the youth wing, chief of the *adat* council, and the respective subvillage heads who are assigned and recruited to guard the forest. Their tasks are to patrol the forest and to detect any threat such as forest fires, illegal logging, fauna hunting, and illegal collection of forest products. The CFGUs report to the district government each case of default found in their respective villages. This has dramatically reduced illegal timber felling. The district government follows up on CFGU complaints by taking necessary action, coordinating forest patrols, and helping to prevent forest fires by coordinating with the central government forestry units in the district.

CFGUs have worked quite well in each *nagari* within Solok District. In Nagari Koto Sani, the unit was able to stop tree cutting for canoe making in the PA. In Nagari Batang Barus, the CFGU has socialized its community members regarding the importance of forest, gotten the users involved in forest patrols, and sent periodic reports to district forestry services. However, while some *nagari*s have taken initiatives to safeguard the PAs in their vicinities, others have not been so successful and hence the forest conditions have varied across the *nagari*s. We therefore examine in the following sections how revitalized *nagari* administration has responded to the decentralization opportunity through an examination of the implementation of policy documents on PAs management, and then by triangulating field observation, interviews with key informants, and a household survey.

14.5.3 Nagari Initiative

The CFGUs from the villages surrounding Singakarak Lake (D-III) and from the rural areas of Solok (D-IV) took additional strategic steps to protect forest. These varied across *nagari*s, from creating forest protection regulations to activities that stopped illegal logging and forest clearing. This is quite an interesting development since the *nagari*s had previously lost control over forest resources after the enactment of the Desa administrative system in 1983.

For example, in Nagari Jawi-Jawi, located in the Solok area of D-IV, *nagari* conservation of forest is achieved by not allowing any further logging and forest clearance for agriculture. It is provided in the regulation that people who are currently farming in the PAs are allowed to continue but no more expansion is allowed. Even though this regulation still needs district government approval, at the local level it has taken effect: no more forest clearance has been carried out and farmers follow the regulation.

In Nagari Guguak Malalo, which is situated in rural areas surrounding Singkarak Lake of D-III, forest regulation is designed for ecological protection, as the villagers want to protect the forest in view of the threat of landslides in the periphery of their villages. Through enactment of this regulation (a) no animal killing and hunting is allowed, (b) no trees are allowed to be cut, and (c) no shifting cultivation is allowed.

Padang Laweh Malalo Village of D-III is quite distinct as far as conservation forest is concerned. The *nagari* administration negotiated with a forestry department agency at the regional level to readjust the PA boundary by removing some water sources from the state conservation forest and putting them under *nagari* control. With this area under community control, the villagers feel that the forest where water sources are found will be better protected as compared to those under state forest regulation.

There have been attempts by local people to stop illegal logging in Nagari Talang of Solok District (D-IV), where the youth have reported cases of illegal logging carried out by ex-police officers and other officials working for the *nagari* government. In Nagari Saningbakar along Singkarak Lake (D-IV), it was reported that, to rehabilitate critical land in forest areas, migrants from this *nagari* mobilized and invested as much as US$153,000 for rehabilitation of critical land.

14.5.4 Varying Results of Multilevel Decentralization and Comanagement of Protected Areas

Our household surveys show that decentralization has had a variable impact on forest management within PAs. In general, there are similar forest-related activities across districts. However in rural areas of Solok District (D-IV) nearly half of the households reported some positive impact with regard to forest management after revitalization of the *nagari* administration system (Table 14.2). In contrast, D-I, D-II, and D-III households revealed a lower impact of *nagari* revitalization in managing forest, meaning that there was very little comanagement. To those who mentioned a change in forest protection since *nagari* revitalization, we asked further questions, such as what activities *nagaris* have been taking with regard to PA. Two main responses were guarding the forest and regulating forest use (Table 14.3). This implies that the communities are already participating in several important aspects of forest conservation, engaging in fact in comanagement. These responses are in line with conservation initiatives taken by the *nagari*. For example Nagari Koto Sani in Solok District has implemented *nagari* regulations regarding forest; that is, (a) villagers are allowed to cut timber for their own use, (b) if timber is for sale within the *nagari*, a tax is levied by the *Nagari* Council at US$5 per m^3, and (c) no timber transportation outside the *nagari* is allowed.

As a consequence of the local initiative, there is a significant change in the number of households involved in illegal logging in D-IV (Table 14.3). The varying degrees of decentralization and comanagement also have effects on local understanding of PA management authority. As shown in Table 14.4, in D-IV local people perceived the *nagari* as having authority over the management of PAs while in the rest of the district local people perceived district and provincial government as having authority over the PAs and not the *nagari*. This implies no decentralization.

14.5.5 Conditions for Comanagement

As this chapter argues, comanagement can only happen when there is a multilevel decentralization; from central to district and down to village level government. Second, comanagement happens only if there is a local institution to cooperate with. Third, there must be a clear incentive for the parties to participate. Our cases show that D-IV district performs comanagement while the rest do not. We explore three conditions for comanagement below.

14.5.5.1 Re-decentralization

Our cases show that there are significant differences in re-decentralization across districts D-I to D-IV. D-I represents no re-decentralization from district to local level institutions while D-IV represents re-decentralization and is performing comanagement. As mentioned earlier, current decentralization policy in Indonesia opens room for creativity in the provision of administration and public services. However, there are differences between districts in their initiatives under the decentralization framework. In many places, however, decentralization ends at district level government.

14.5.5.2 Local Institutions

Comanagement is only feasible with the active involvement of local institutions. In many parts of the world during the last four decades, there have been attempts to replace various local institutions with centrally designed homogenous administrative models. These attempts have further weakened local institutions. Because comanagement is not possible through weak local institutions, there is a need to revitalize and empower the local institutions.

Revitalization of traditional local institutions like the *nagari* in West Sumatra is an appropriate way to re-decentralize. As mentioned earlier, the *nagari* is a local, traditional

TABLE 14.2 Number of Households Involved in Forest-Related Activities

Level of Decentralization and Comanagement	Total Household Sample	Number of Households						
		Involved in Forest Farming	Collecting Firewood	Hunting and Trapping Animal/Birds	Collecting NTFP	Possessing Chain Saw	Currently Involved in Timber	Previously Involved in Timber Felling
D-I (no revitalization)	70 (100.00)	34 (48.6)	40 (57.1)	6 (8.6)	8 (11.4)	0 (.0)	15 (21.4)	17 (24.3)
D-II (revitalization but no decentralization	75 (100.0)	45 (60.0)	58 (77.3)	17 (22.7)	13 (17.3)	3 (4.0)	17 (22.7)	25 (33.3)
D-III (revitalization, decentralization, but no collaboration)	74 (100.0)	60 (81.1)	52 (70.3)	16 (21.6)	7 (9.5)	4 (5.4)	21 (28.4)	34 (45.9)
D-IV (revitalization, decentralization, and collaboration)	80 (100.00)	54 (67.5)	61 (76.3)	11 (13.8)	21 (26.3)	0 (.0)	3 (3.8)	37 (46.3)
Total	299 (100.0)	193 (64.5)	211 (70.6)	50 (16.7)	49 (16.4)	7 (2.3)	56 (18.7)	113 (37.8)

Note: Figures in parenthesis indicate percentages.

III. SOCIOECOLOGICAL SYSTEMS AND NEW FORMS OF GOVERNANCE

TABLE 14.3   Perceived Impact of Decentralization and Comanagement of Protected Areas

Level of Decentralization and Comanagement	Total Household Sample	Perceived Impact of *Nagari* Revitalization on Protected Area Management			*Nagari* Activities in Managing Protected Area	
		No Impact	Do Not Know	Positive Impact	Regulating Forest Use	Guarding Forest
D-I (no revitalization of village institution)	70 (100.0)	69 (98.6)	1 (1.4)	0 (0)	0 (0)	0 (0)
D-II (revitalization without decentralization	75 (100.0)	39 (52.0)	25 (33.3)	11 (14.7)	4 (5.3)	3 (4.0)
D-III (revitalization, decentralization, without collaboration)	74 (100.0)	52 (70.3)	17 (23.0)	5 (6.8)	0 (0)	0 (0)
D-IV (revitalization, decentralization and collaboration)	80 (100.0)	26 (32.5)	16 (20.0)	38 (47.5)	38 (47.5)	19 (23.8)
Total	299 (100.0)	186 (62.2)	59 (19.7)	54 (18.1)	42 (14.0)	22 (7.4)

Note: Figures in parenthesis indicate percentages.

TABLE 14.4   Perceived Existence of Protected Areas, Authority Over Protected Areas

Level of Decentralization and Comanagement	Total Household Sample	Perceived Existence of Protected Area			Perceived Authority Over Protected Areas		
		No	Exist	Do Not Know	Government	*Nagari*	Other
D-I (no revitalization)	70 (100.0)	20 (28.6)	42 (60.0)	8 (11.4)	25.00 (35.71)	1 (1.43)	44.00 (62.86)
D-II (revitalization but no decentralization	75 (100.0)	8 (10.7)	55 (73.3)	12 (16.0)	38.00 (50.67)	3 (4.00)	34.00 (45.33)
D-III (revitalization, decentralization, but no collaboration)	74 (100.0)	8 (10.8)	58 (78.4)	8 (10.8)	30.00 (40.54)	4 (5.41)	40.00 (54.05)
D-IV (revitalization, decentralization, and collaboration)	80 (100.0)	14 (17.5)	57 (71.3)	9 (11.3)	16.00 (20.00)	21 (26.25)	43.00 (53.75)
Total	299 (100.0)	50 (16.7)	212 (70.9)	37 (12.4)	109 (36.40)	29 (9.70)	161 (53.90)

Note: Figures in parenthesis indicate percentages.

sociocultural and political unit that has a strong local basis as compared to the *desa* administrative system introduced during the New Order by the central government.

14.5.5.3 Local Incentives

The provision of incentives at the local level is a real challenge for local government seeking to get local communities involved in PA management. If this could be attained, dependence on external resources to finance PAs management could be reduced. In order to attain this, the valuation of benefits from PAs has to be expanded beyond biodiversity conservation to ecological and environmental services (see for example Sunderlin et al., 2005; Smith and Scheer, 2003). Sunderlin et al. (2005) have identified four types of direct payments for forest dominant environmental services, namely, carbon, hydrological protection, biodiversity conservation, and recreational values. By securing payment in return for these services, the finance problem of PAs could be solved. A watershed protection fee, for example, is a potential source of finance for PAs (see Spergel, 2002).

These last two developments suggest a different approach for PAs management and to the role each stakeholder could play. On the management level, Sayer (2000) suggests that a conservation agency could adopt output-based systems ensuring effective collaboration among all stakeholders. Hence, the opposing objectives of local people, including local government, and the establishment of a PA could lead to a movement for reconciliation. However, for many local governments in developing countries, it is hard to imagine environmental services that generate local revenue. Their basic questions will go like this: who is going to pay for a carbon sink, how are we to increase the low revenue derived from recreational value, and what sort of value is to be given to biodiversity conservation? The only clear ecological benefits would probably be hydrological protection, but again who is going to pay for this service?

With current decentralization and the financial balance between central and district government in Indonesia, a portion of the natural resources tax is returned to provincial and district government. This is a significant change in the sharing mechanism of collected tax under the decentralization law. Earlier, Indonesia was known for the most centralized taxation system in the world (Simanjuntak, 2001).

Surface water for a hydropower plant is a taxable natural resource. The National Power Corporation (PLN) as the operator of Singkarak hydroelectric power plant (HEPP) pays an amount of IDR 1.8 billion (US$180,000) per year as a surface water tax (*Mimbar Minang*, January 14, 2003). According to Law No. 34/2000 on tax and regional redistribution, 70% of the water tax should be returned to district governments. District governments, in turn, should allocate 10% of the amount to village level government. Surface water tax received by the district government from Singkarak HEPP creates an incentives mechanism to get communities involved in PAs management. Using this money, the district government also persuades the *nagari* government to protect forest, by such measures as issuing *nagari* regulations on forest protection and financing the operation of a *nagari* forest guard task force. Solok District (D-IV) government has enjoyed this tax return since the implementation of government decentralization in 2001.

Direct surface water tax or a watershed protection fee at a global level is not exclusive to Solok, however. In Columbia, Spergel (2002) reports, the 1993 Environmental Law required hydroelectric plants to transfer 3% of their revenues to regional governments (and an additional 3% to municipal governments) to carry out watershed conservation projects and urban sanitation projects. In Quito, Ecuador, water consumers pay a small surcharge on their monthly water bills to finance the cost of maintaining the forest cover of the watershed that supplies the city with drinking water. In Laos, Spergel reports, the developers of the proposed $1.3 billion Nam Theun hydroelectric dam have agreed to pay $1 billion per year for 30 years into a "watershed conservation fund" to protect the pristine forests and endangered wildlife on the steep mountain slopes above the dam. Conserving the forests is also way of preventing the dam from silting up, thereby extending the dam's economic life by more than 50% (Spergel, 2002: 369).

14.6 PROPOSED COMANAGEMENT MODEL FOR PROTECTED AREAS

A review of several cases around the world and our field study show that it is not easy for local government to take an active role in PAs management. The basic reason is that local administration and communities do not see any direct benefit in terms of local revenue from protecting conservation forest. In addition, the central government has been reluctant to involve local communities in PAs management, especially of conservation forest. De jure all conservation forest is under the authority of the central government, and protection forest management has been devolved to local government. But, if a clear incentive is available, local government is willing to share responsibility. This has implications for the comanagement model. A review of our findings shows that the following definition of comanagement by Borrini-Feyerabend et al. is equally applicable in Indonesia:

> a situation in which two or more social actors negotiate, define and guarantee amongst themselves a fair sharing of the management functions, entitlements and responsibilities for a given territory, area or set of natural resources. *Borrini-Feyerabend et al. (2000: 1)*

This definition implies a need to modify current PAs management practices under government decentralization in Indonesia and other developing countries.

Our proposed framework for comanagement takes several factors into consideration including the adoption of a decentralization policy in administration and at the financial level. Furthermore, due importance must be given to local people's participation in PAs management by revitalizing local institutions in forest management.

The three main pillars for PA comanagement in Indonesia are regional autonomy, forestry management, and financial autonomy. A regional autonomy law gives local autonomy in the adoption of locally appropriate administration, without the need to follow a homogeneous standardized national model, as it was under the centralized model of village administration. In Indonesia, where multicultural identities exist, adopting a single model of village government has proven inappropriate. Hence, current decentralization by local government could adopt a more appropriate structure according to local culture and tradition. In West Sumatra province, local government has adopted the "return to *nagari*" concept, which means revitalizing local village level governance. The current Forestry Law, Law No 41/1999 acknowledges the importance of the local government and the local community getting involved in PAs management. The central government has devolved power to district level government to manage protection forest. Financial incentives and burden-sharing mechanisms provide

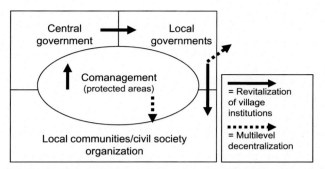

FIG. 14.2 Proposed comanagement model for protected areas in Indonesia. *Adopted from World Bank, 1999. Report from the International CBNRM Workshop, Washington DC, 10–14 May 1998. URL: http://www.worldbank.org/wbi/conatrem/.*

for local government to receive intergovernmental transfers of funds to provide basic services at the district level and to equip regional autonomy with a right to use money from external sources through debt or grant.

Based on these considerations we propose a comanagement model for PAs by devolving power to village level government such as the *nagari* in West Sumatra, and similar local institutions in other parts of the country. Our proposed model is modified from the World Bank comanagement model by adding re-decentralization and revitalization of local institutions as preconditions for comanagement. In addition, we also include local incentive structures in the framework. Our overall framework is presented in Fig. 14.2.

Finally, to not treat the adoption of a comanagement model as yet another panacea, outsiders who extract resources from PAs should be subjected to the jurisdiction of the local people. These local stakeholders should be able to take action against such outsiders. In this situation, comanagement can be effective against external intruders. But it can also be captured by local elites for their individual benefit. Hence empowerment of the local community through the provision of opportunities to sue both external intruders and local elites should be pursued so that neither of these groups under the cover of local representation and participation becomes yet another source of exploitation.

Acknowledgements

An earlier version of this chapter was published by the *Journal of Legal Pluralism*. "Yonariza and Ganesh Shivakoti., 2008. Decentralization and co-management of protected areas in Indonesia. *J. Legal Pluralism*. No. 57(1):141–165," which is duly acknowledged.

References

Aden, J., 2001. Decentralization of Natural resource sectors in Indonesia: opportunities and risks. Eases discussion paper series September 2001. The World Bank East Asian Environmental and Social Development Unit, Washington DC. Available on-line: http://www.gtzsfdm.or.id/documents/dec_ind/gv_pa_doc/Decent_of%20 Natural.pdf [accessed: 3/31/2004]

Andersson, K.P., Clark, C.G., 2004. Decentralization reforms: help or hindrance to forest conservation? In: Paper presented at the Conference of the International Association for the Study of Common Property (IASCP) in Oaxaca, Mexico, August 9–13.

Berkes, F., George, P., and Preston, R., 1991. Co-management: The evolution of the theory and practice of joint administration of living resources. Paper presented at the Second Annual Meeting of IASCP University of Manitoba, Winnipeg, Canada, Sept. 26–29, 1991.

Barber, C.V., Talbott, K., 2003. The chainsaw and the gun: the role of the military in deforesting Indonesia. J. Sustain. Forestry 16, 137–166.

von Benda-Beckmann, F., von Benda-Beckmann, K., 2001. Recreating the Nagari: decentralisation in West Sumatra. Max Planck Institute for Social Anthropology Working Paper No. 31. Max Planck Institute for Social Anthropology, Halle.

Borrini-Feyerabend, G., Farvar, M.T., Nguinguiri, J.C., Ndangang, V.A., 2000. Co-management of natural resources: organising, negotiating and learning-by-doing. Heidelberg, GTZ and IUCN, Kasparek Verlag. http://nrm.massey.ac.nz/changelinks/cmnr.html.

Carlsson, L., Berkes, F., 2005. Co-management: concepts and methodological implications. J. Environ. Manage. 75, 65–76.

Casson, A., Obidzinski, K., 2002. From new order to regional autonomy: shifting dynamics of "illegal" logging in Kalimantan, Indonesia. World Dev. 30, 2133–2151.

Clifton, J., 2003. Prospects for comanagement in Indonesia's marine protected areas. Mar. Policy 27, 389–395.

Da Silva, P.P., 2004. From common property to comanagement: lessons from Brazil's first maritime extractive reserve. Mar. Policy 28, 419–428.

Dewi, S., Belcher, B., Puntodewo, A., 2005. Village economic opportunity, forest dependence, and rural livelihoods in East Kalimantan, Indonesia. World Dev. 33, 1419–1434.

Elliott, C., 1996. Paradigms of forest conservation. Unasylva 47. No. 187.

FWI/GFW (Forest Watch Indonesia and Global Forest Watch), 2002. The State of the Forest: Indonesia. Bogor, Indonesia: Forest Watch Indonesia, and Washington DC: Global Forest Watch.

Gibson, C.C., Williams, J.T., Ostrom, E., 2005. Local enforcement and better forests. World Dev. 33, 273–284.

Griffiths, M., Van Schaik, C., Rijksen, H.D., 2002. Conserving the Leuser ecosystem: politics, policies, and people. In: Terborgh, J., van Schaik, C., Devenport, L., Rao, M. (Eds.), Making Parks Work, Strategies for Preserving Tropical Nature. Island Press, Washington, Covelo, London, pp. 203–217.

Haeruman, H.J., 2001. Financing integrated sustainable forest and protected areas management in Indonesia: Alternative mechanisms to finance participatory forest and protected areas management. In: Paper presented at the International Workshop of Experts on Financing Sustainable Forest Management, Oslo, Norway, 22–25 January IPF/IFF/UNFF Processes.

Hiller, M.A., Jarvis, B.C., Lisa, H., Paulson, L.J., Pollard, E.H.B., Stanley, S.A., 2004. Recent trends in illegal logging and brief discussion of their cause: a case study from Gunung Palung National Park, Indonesia. J. Sustain. Forestry 19, 181–212.

IUCN (International Union for Conservation of Nature), 1994. Guidelines for Protected Area Management Categories. IUCN World Commission on Protected Areas with the assistance of the World Conservation Monitoring Centre, Gland, Switzerland and Cambridge, UK.

Laurance, W.F., 2004. The perils of payoff: corruption as a threat to global biodiversity. Trends Ecol. Evol. 19, 399–401.

Lindayanti, R., 2002. Shaping local forest tenure in national politics. In: Dolsak, N., Ostrom, E. (Eds.), The Commons in the New Millennium. MIT Press, Cambridge, Mass, pp. 221–264.

Lutz, E., Caldecott, J., 1996. Decentralization and biodiversity conservation. A World Bank Symposium. World Bank, Washington DC (downloaded: 3/29/2004).

Mccarthy, J.F., 2000. Wild logging: the rise and fall of logging networks and biodiversity conservation projects on Sumatra's rainforest frontier. CIFOR Occasional Paper No. 31. Center for International Forestry Research, Bogor. Available online: http://www.cifor.cgiar.org/publications/pdf_files/OccPapers/OP-31.pdf [accessed: 3/29/2004].

Mccarthy, J.F., 2002. Power and interest on Sumatra's rainforest frontier: clientelist coalitions, illegal logging and conservation in the Alas Valley. J. Southeast Asian Studies 33, 77–106.

Mccarthy, J.F., 2004. Changing to gray area: decentralization and the emergence of volatile socio-legal configurations in Central Kalimantan, Indonesia. World Dev. 32, 1199–1233.

Mimbar Minang., 2003. "Demo menuntut pembukaan pintu air." Mimbar Minang, 14 January.

Obidzinski, K., 2004. Illegal logging and the fate of Indonesia's forests in times of regional autonomy. In: Paper for the Panel "Nontrivial Pursuits: Logging, Profits and Politics in Local Forest Practices in Indonesia" presented at the Conference of the International Association for the Study of Common Property (IASCP) in Oaxaca, Mexico, August 9–13 (draft).

Obidzinski, K., Barr, C., 2003. The effects of decentralization on forests and forest industries in Berau District, East Kalimantan. CIFOR's Case Studies on Decentralisation and Forests in Indonesia, Case Study No. 9. Center for International Forestry Research, Bogor. Available on-line: http://www.cifor.cgiar.org/publications/pdf_files/Books/Decentralisation-Case9.pdf [accessed: 3/29/2004].

Rao, K., 2001. Lessons from global experience in protected areas planning and management. In: Paper presented at Review of protected areas and their role in socio-economic development in the four countries of the lower Mekong River region, Vietnam First National Round table, 14 September.

Ravenel, R.M., 2004. Community-based logging and defacto decentralization: illegal logging in the Gunung Palung area of West Kalimantan, Indonesia. J. Sustain. Forestry 19, 213–237.

Rhee, S., 2000. De facto decentralization during a period of transition in East Kalimantan. Asia-Pacific Community Forestry Newsletter 13: 34–40. Available on-line http://www.recoftc.org/documents/APCF_Newsletter/13_2/Defacto_rhee.pdf [accessed: 3/30/2004].

Sayer, J.A., 2000. Forest protected areas: time is running out. In: Rana, D.S., Edelman, L. (Eds.), The design and management of forest protected areas. Papers presented at the Beyond the Trees Conference 8–11 May 2000, Bangkok, Thailand. 1–10.

Simanjuntak, R., 2001. Local taxation policy in the decentralizing era. In: Paper presented at Domestic Trade, Decentralization and Globalization: A One Day Conference, Hotel Borobudur Jakarta, 3 April.

Smith, J., Scheer, S.J., 2003. Capturing the value of forests carbon for local livelihoods. World Dev. 31, 2143–2160.

Spergel, B., 2002. Financing protected areas. In: Terborgh, J., van Schaik, C., Devenport, L., Rao, M. (Eds.), Making Parks Work, Strategies for Preserving Tropical Nature. Island Press, Washington, Covelo, London, pp. 363–382.

Sudana, M., 2004. Winners take all. Understanding forest conflict in the era of decentralization in Indonesia. In: Paper presented at the Conference of the International Association for the Study of Common Property (IASCP) in Oaxaca, Mexico, August 9–13.

Sunderlin, W.D., Angelsen, A., Belcher, B., Burgers, P., Nasi, R., Santoso, L., Wunder, S., 2005. Livelihoods, forests, and conservation in developing countries: an overview. World Dev. 33, 1383–1402.

World Bank, 1999. Report from the International CBNRM Workshop, Washington DC, 10–14 May 1998. URL: http://www.worldbank.org/wbi/conatrem/.

Wyckoff-Baird, B., Kaus, A., Christen, C., Keck, M., 2000. Shifting the power: decentralization and biodiversity conservation. Washington, DC, Biodiversity Support Program (BSP).

Further Reading

Legislation of the Republic of Indonesia.
Conservation Act No. 5/1990.
Financial Balance Act No. 33/2004.
Forest Act No. 41/1999.
Government Regulation No. 68/1998 on Nature Reserve and Nature conservation.
Government Regulation No. 34/2002.
Government Regulation No. 45/2004 on Forest Protection.
Local Government Act No. 22/1999.
Regional Autonomy Act No. 32/2004.

Dynamism of Deforestation and Forest Degradation in Indonesia With Implications for REDD+

S. Sharma

WWF-Nepal, Kathmandu, Nepal

15.1 INTRODUCTION

Reducing Emissions from Deforestation and Forest Degradation (REDD+) is the mechanism developed to offer incentives for reducing greenhouse gases (GHGs) while significantly adopting forest management approaches that are sustainable. This mechanism was first agreed at the 13th Conference of the Parties (COP 13) and, subsequently, the Indonesian government committed to employing this arrangement. The government would attain this objective through reducing emissions of GHGs by 26% if "business is as usual" through its own support; while 40% with international support, respectively (Mahdi et al., 2015).

The literature shows that deforestation is alarmingly high, especially at Sumatra and Kalimantan where deforestation accounts for 2.7–1.3%, respectively, for higher timber value (Indrarto et al., 2012). According to Abood et al. (2014) "out of 14.7 million hectares of destroyed forest in Indonesia in between 2000 and 2010, 12.8% of forest was removed for fiber plantations, 12.5% for logging and 6.8% for palm-oil plantations; the remaining was removed through businesses and mining." Prominently, 45% of forest loss has arisen on industry leased lands (ibid). Fiber- and timber-based industries on the one hand and futile efforts in regeneration of logged area on the other hand has resulted in considerable forest degradation in many parts of Indonesia (Wardojo and Masripattin, 2002).

Deforestation is linked to Indonesian finance and also shaped the political economy coupled with the institutional landscape, while overall affecting the forestry sector; explicitly, REDD+ (Indrarto et al., 2012). Government has realized this complexity, which is prominent through a number of reforms in national and sectorial polices (Wardojo and Masripatiin, 2002) mostly trying to find a balance between allocation of forest cover and agricultural land (Galudra and Sirait, 2006). However, the actual implementation of policy issued by the government is significantly low, creating a huge gap between policy and practice (Yonariza et al., 2015). Specifically, with REDD+ implementation, Indonesia's national polices on emissions reduction are not included or not yet becoming mainstream policy in long-term national plans (Mahdi et al., 2015).

This intricacy of trying to balance the economy and reduce deforestation in the face of REDD+ creates a situation of complexity whereby the needs and problems of forest-dependent people may be lost. This is because either polices are ad hoc to pursue space at the international REDD+ arena and further impose mandatory actions against local forest-dependent people to use and manage the resources to implement decisions made globally. Because the forestry sector combined with the wood manufacturing, pulp, and paper industry contributes to US$21 billion to the country's GDP (ITS Global, 2011), there is at least the possibility that limiting production from the forestry sector could be anticipated. To overcome the commitment made in the international arena, Indonesia would be inevitable to decrease emissions through declaring forests as protected area.

In this current backdrop, the government of Indonesia is planning to implement REDD+ without acknowledging the fact that forest management is a series of socioecological interactions that involves institutions, political pressures, and users actions that have come up through experiences and procedures. The accomplishments of REDD+ are incredible if natives are not involved in the implementation process. REDD+ is directed to benefit forest and native people, yet has uncertainty over poorly demarcated forest tenure and appropriate incentives agreed upon by the consensus building process. In addition, the unwillingness of government to devolve to communities in the past also engenders fear that REDD+ could decentralize well-functioning community forest (Phelps et al., 2010).

In this regard, this chapter tries to provide an overall picture of Indonesian forestry to understand the background of deforestation and degradation in which REDD+ polices and process emerge. This is important because the policy formulation process in Indonesia is usually done without in-depth homework on understanding the grassroots problems and consequences often created by huge policy practice gaps and situations of conflicts aroused between government and communities (Yonariza et al., 2015; Mahdi et al., 2015). This chapter will aid decision makers, donors, and practitioners on prospects and encounters in implementing a REDD+ instrument to support an evidence-based REDD+ decision-making process. This study is part of the project "Bridging Policy Practice Gap in the Effective Implementation of REDD+ Programs in SE Asia: Collaborative Learning among Indonesia, Thailand and Vietnam," with financial and technical support from the Toyota Foundation from November 2013 to November 2014.

15.2 DEFORESTATION AND FOREST DEGRADATION IN INDONESIA

Indonesia is inhabited by 30 million individuals dependent on forests for their livelihoods (Sunderlin et al., 2000). Indonesia's forest cover spread at 123,459,515 and 133,694,685.18 ha in 2005 and 2008, respectively. Although forest cover in Indonesia at first appears to have increased, these changes involve discrepancies aroused due to the different methods in information gathering used. The differences in forest types, forest classifications, and variations in data collection and analysis have led to inconsistencies in forest data in Indonesia; therefore, the authentic statistics on Indonesian's forest status are difficult to assemble. Forestry law in Indonesia divides forest into production forest, protection forest, and conservation forest. As the title put forward, conservation and protection forest is for protecting ecosystem functioning while production forest is mainly for economical purposes further divided into permanent production, limited production, and convertible production (ITS Global, 2011). Yet, activities like mining may be conducted at any forest type through lease use permits issued by the Ministry of Forestry (MoF). This endowment, too, has controlled conceivable deforestation in Indonesia. Forest cover variations perhaps may be the result of forest degradation coupled with planned and unplanned deforestation, which may be both legal and illegal. Government planned changes for estate and agricultural crops are legitimate under the law. Unplanned or illegal logging arouses degradation, which leaves forest areas additionally exposed to deforestation as depredated forests are calm to clear. Large forest fires associated with agricultural undertakings are commonly not shadowed by natural generations, further aggravating the natural ecosystem.

Broadly, forest losses have occurred when the area is allocated for other purposes. Transformations of forest areas to palm states have been triggered by high crude oil. As per the Ministry of Agriculture, biofuel and oil palm plantations in Indonesia have touched 7,007,867 and 8,430,026 million ha in 2008 and 2010, respectively. According to the Indonesia National Biofuel Team, to bump into the national target, 10.25 million ha of plantations would be required. Similarly, mining and infrastructure development have converted a large amount of forest into development activities. Releasing of grants for commercial timber plantations is another form of deforestation further aggravated by disobeying permits; chopping trees more than designated and short of permitted targets.

Also to be deliberated is the local community's involvement in forest degradation through expansion of agriculture and reduction of 4 million ha of forest area between 1985 and 1997 through swidden agriculture. Conversely, this system of agriculture is being used less in areas of Kalimantan because communities currently prefer planting productive trees, which they found not only profitable but realized continuous plantation would strengthen their informal land rights. Growth of a forest development economy followed by strong timber market demand and high estate crop value has contributed to acceleration of deforestation and forest degradation. It was also found that clarity on tenure rights and resource boundaries have also augmented the progression. During 2007, the MoF quantified that 52.60% of settlements in Indonesia are positioned inside forested area, which is hitherto to be decided and created many conflict situations.

Although the government is inclined toward the implementation of REDD+, the complex deforestation and forest degradation situation may condense the ability of Indonesia's forest to allay carbon. Additionally, Indonesia has more than 21 million ha of GHG emitting peatlands that in 2000 to 2006 emitted an average of 903,000 Gg of CO_2 annually. Additional estimates conducted by the National Climate Change Council indicated Indonesia's 2005 carbon emission level was "2.1 Gt CO_{2e}, growing 1.9% yearly to 2.5 Gt CO_{2e} in 2020 and 3.3 Gt CO_{2e} in 2030" (Rusli, 2011). These facts show that Indonesia must make contributions toward reducing these emissions. In the process, the government has finalized guidelines to achieving the emissions commitment through reforestation, plantation under community forest, and development of *Hutan Tanaman Rakyat* (HTR). To propose justifiable community-based sustainable plantation forest management HTR was familiarized by the state inquiry lead by Verchot et al. (2010) deliberates that REDD+ involvement with HTR is likely to fight to achieve anticipated outcomes. This is because it is operative to manage deforestation in the existing forest or otherwise to move plantation to new areas.

15.3 INSTITUTIONAL CHARACTERISTICS OF DEFORESTATION AND FOREST DEGRADATION

Indonesia has a huge span of tropical forest and has an immense role in accordance with international forums. The United Nations Forum on Forests (UNFF) is one of the international conventions that Indonesia is assigned to. With this agreement, governments have introduced priority policies in forestry, reducing illegal logging and forest fires, rehabilitating deforested forest, and finally decentralized management. Surrounded by this arrangement a social forestry approach was introduced to ensure sustainable forest and community prosperity together. Indonesia has also sanctioned on Convention of Biological Diversity (CBD) for achieving sustainable protection of biodiversity through a National Biodiversity Strategy and Action Plan (NBSAP). But, this was more focused on compiling biodiversity information otherwise to reduce deforestation and degradation. Subsequently, the Convention on International Trade in Endangered Species of Wild Fauna and Flora (CITES) was ratified by the government, and the MoF was given the managing authority. On the contrary, Indonesia is yet to manage most of its obligations. This is because officials either did not understand the conventions or could not manage their time accordingly (Soehartono and Mardiastuti, 2002). To reduce illegal logging in tropical forest sectors, Indonesia became involved in Forest Law Enforcement, Governance, and Trade (FLEGT). Likewise, the global economy shapes Indonesia's polices. The World Trade Organization (WTO) and the International Monetary Fund (IMF) are the major actors in establishing laws in Indonesia. This was prominent during the years 1998–2003, while economic restructuring in Indonesian forest was done by the IMF, whereby it compelled lowering taxes on timber logs and rattan, decreasing regulations on plywood marketing, and reducing natural forest conversions.

A predominant reason for deforestation and forest degradation is feeble forest administration, which is also affected by frameworks affecting other sectors. According to Indarto et al. (2012), "Law No. 41 on Forestry, Law No. 22 on Regional Governance (often called the Regional Autonomy Law; later replaced by Law No. 32/2004), Law No. 18/2004 on Estate Crops, Law No. 4/2009 on Mineral and Coal Mining, Law No. 32 on the Environment and Law No. 26/2007 on Spatial Planning" are such laws directly affecting forest management. These regulations are erratic and generate confusion due to overlapping and imprecise authorities. For instance, the legislature provides authority to regional heads to grant plantation crop permits. This creates confusion in law enforcement and poses unclear stakes of state forest to deal with deforestations. Irrespective of legislature slipups, instances of the MoF replying contradictorily in permit application have become apparent. Cases happened where the MoF should have taken actions besides managerial violations, but then again the canceled permits were "whitewashed" and reissued.

Forest management suffers from basic and diverse drawbacks in permit licensing rules, law enforcement, and capacity. For REDD+ to be successful or deforestation and degradation to be reduced, forest management should be effective at each level. In the context where multiple complexities are involved, there is no guarantee of REDD+ pursuing fundamental changes. The most vital transformation for regulation is to incorporate the ideologies of good governance in the process of issuing timber licenses and permits. Possibilities exist for incorporating strict legislation on permit mechanisms, controlling responsibilities, and enhancing sanction mechanisms typically for guilty officers and personnel. The accountabilities of regional and central level authorities needs clarification so that processes will be inclined toward obtaining sustainable forestry with clear and transparent monitoring and supervision instruments as a prerequisite. Detailed bureaucratic restructuring with methodical efforts to eliminate corruption is vital. If these procedures are far from being carried out, law enforcement at ground level would be difficult to accomplish. Disappointment to achieve these results will concomitantly compromise endeavors on reducing deforestation and degradation. Consequently, benefits from REDD+ would be overshadowed by negative effects of reducing forest access to forest-dependent communities. REDD+ is not only related to revenue generation and sharing, but is tied closely to forest governance.

15.3.1 Deforestation in the Context of Economic Policy

The government in 2008 released tax reduction provisions in forestry sectors specifically for pulp, paper, metals, and other forest products. Similarly, regulations on "tariffs on non-tax state revenue" permitted open mining within protection forest, which clearly shows the government's concern over economic growth. At about the same time, a package on biofuel development was initiated in 2006 as a provision to promote biofuels. This attracted a lot of attention from entrepreneurs, and agricultural land was soon converted to growing palm, cassava, and castor oil.

Indonesia is dependent on natural resources for economic development. In the process, timbers are extracted from the forest while the process of timber plantations in new areas with palm and oil industries is rampant, seeking to secure government's earning. The mining sector not only increases revenue to the government but at the same time causes massive deforestation and degradation. Despite the fact that this sector contributes only 2.4% of total revenue generated, there are other economic dimensions such as paper, pulp, and mines generating more income. A larger market for industrial crops has promoted expansion for such monocultures. Rubber and palm oil have experienced significant increases and this concept indicates the state's devotion to a sectorial attitude. This signifies that the plantations that made a lot of revenue should be producing at certain quantity levels and specifications. Subsequently, the government is likely to use possible resources to generate revenue, leaving behind considerations for biodiversity and land-use changes.

REDD+ needs to be applied carefully and judgmentally. Higher market demand for estate plantation and the mining business has influenced the political and economic dimensions of Indonesia. The government believes REDD+ is an effective way to mitigate carbon and render huge benefit, but it is sometimes skeptical about the numerous associated forestry issues. No possible measures for clarifications on REDD+ solving issues exists as there are too many overlying regulations and processes. The country is also not free from corruption and lacks transparency (Dermawan et al., 2011). In a country like Indonesia with an economy that endures by relying on its natural resources, improvements in good governance and corruption-free implementation and understanding is essential.

15.3.2 Decentralization and Benefit Distribution With Effects for REDD+

The Indonesian government began a decentralization mechanism in accordance with devolution mechanisms around the globe. To gauge progress, Government Regulation No. 62/1998 was released, which delegated forest governance from the central to the regional government, which was then delegated to districts to carrying out forest, soil, and water rehabilitation and reforestation. Although the central government intended to mainstream decentralization in the forestry sector, it conversely added a financial and managerial burden to regional heads. The thrust for decentralization began following the demands and conflicts of communal resources and underprivileged benefits. Later, the government delivered regulations that provided governor and district heads with the privilege to issue timber permits up to 10,000 and 100 ha, correspondingly. This policy was later reinforced by allotting forest product withdrawal permits and community involvement in forest innovativeness. The basic aim of this policy was to involve local communities in managing resources while securing economic remuneration accordingly. Also, a number of laws were passed that were targeted to balance power between the regional and central governments, by pursuit of which district and municipal heads granted forest enterprises permits in various arrangements.

Notwithstanding the permits envisioned for small-scale household operations, loggings were done with weighty equipment for areas greater than 100 ha. The same firms were granted permits numerous times without considering ecological and rotational logging ideologies. This differed from centrally controlled permit provisions, which keep an eye on Indonesian Selective Cutting and Planting (ISCP) principles. Though most of the procedure undertaken at the district and municipal level did not adhere with ISCP principles, the policy was put in place. The result showed massive deforestations and forest degradations. This policy did not allow the use of heavy machinery to carry out logging undertakings. Timber corporations harvested logs in many different ways and could take advantage of diverse forms but communities had generated less profit through this policy. Successively, the ambition of this policy for individual communities to enjoy the forest was not wholly used.

The problem becomes more complex when related to a central, regional, and local context on the verge of decentralization. With the authority set, responsibility for other forest affecting sectors will be geared up with the view of increasing regional revenue. It is likely that in most cases, regional governments will issue estate and mining grants locally without gathering one from the central government; approximately 5–7 million of such permits have been allocated. The minimal capacities of government to act on issues such as this has led to this increasingly widespread drill. The added complications in decentralization in Indonesia are as follows:

a. Lack of timely resource boundary demarcation.

This imposes ambiguities and inconsistencies in polices that have arisen on interpreting what constitutes the boundary of the forest. Forest Law confers state forest to be legalized if it is listed and the process includes designating the area, setting boundaries, mapping the forest, and then establishment. Because the central government carries out boundary mapping, untimely resource boundary demarcation may lead the district government to not issue a permit for an area.

b. Lack of executives at the lowest level.

Forest management is fully the responsibility of the government. Hence the MoF must work with diverse participants and parties at both national and local levels to attain sustainable forest management (SFM). Conversely, there is a lack of executives to manage these forests at the lowest level respective to national parks. As such, there have been incidences of encroachments and rules violations with few of these being formally reported to the central government.

c. Corruption in issuing permits.

The process for issuing permits and distributing revenues is corrupted and contravenes rules and legislation not only in issuing permits but also occurs during supervision. Exacerbating corruption is a lack of clear penalties for abuse of authority. Lack of transparent permit issuance has also facilitated the process of corruption. For instance, prior to issuing permits, parties are required to undergo an environmental impact assessment. But most of the time this is either not properly done or just copied from another region.

d. Weak supervision of granted concessions.

Indonesia lacks the relative ability to monitor resources and procedures. To deal with this, the government is compelled to depend on unverified reports submitted by the concession companies. In a particular case, the companies can fell more trees than permitted.

Participation is important for effective service delivery and supervision, which are not implemented properly in Indonesia. There are inconsistent provisions in policies to guarantee community participation. Forestry law provides a very general agenda on participation while policies on spatial planning conversely have better guarantees on planning and permitting issuance of the forest. However, there have been incidences whereby community advocacy has genuinely influenced the policymaking process. Article 19(2) of Regulation No. 44/2004 confers the boundary delineation phase covers inventorying happenings and third party rights resolution together with the state forest's boundary. This is the formal forum provided to the customary communities to assert their rights. Conversely, this process does not accompany provisions for local participation in the planning mechanism with no penalties involved if it failed to involve communities. For customary communities to claim forest rights, complex procedures and document ought to be followed, which adds further difficulty. The literature shows that no forests have received such customary forest claims, so areas receiving status as village forests are usually denoted as to their customary lands.

As number of complexities of concern to decentralization have been aroused, including discordance among regional and central polices; policy urgencies highlighting natural resources-based development; and increased threats to forests from infrastructure development and benefit-sharing mechanisms from the forest. This has implications for successful implementation of REDD+. Given that regional autonomy is devolved at the local level, it will be difficult to assist management interventions in the state forest at a national scale. On the verge of REDD+ supervision is required from the central government level to commence good law enforcement and good governance.

15.3.3 Right to Forest, Right to Carbon

The state has recognized the indigenous community and their rights but these are often neglected at the national level. It is important to note that Indonesia has not approved the Indigenous and Tribal People's Convention under the International Labor Organization;

however, the government has ratified a UN declaration on indigenous people's rights. This is a nonbinding contrivance that requests countries to show positive pledges to indigenous people's rights.

Nevertheless, community rights are well-accepted in Indonesia's legal framework and especially in the constitution (Article 18 B-2), which "recognizes and respects customary rights and laws as long as they live and solidarity with social advances and mandates of the country." The constitution additionally stresses "respecting traditional cultural identity in accordance to development and civilization." These rights deliver authentic standing to withstand petitions in the court, which is similar to any other citizens of Indonesia. Likewise, the government has provisions to implement customary land rights that are in accordance with the interests of the Indonesian people and the state. The determination of native minority and conflict resolution in regards to indigenous community government has provisions under the *Badan Pertanahan Nasional* (National Land Agency) (BPN). Nevertheless, these specificities do not regulate the state's responsibility to recognize customary law and are often abandoned for technical explanations (Sumardjono, 2008). The forestry law acknowledges the presence of customary forest until and unless it contradicts with the state forest and, most of the time, it is still measured as a state resource (Safitri, 2010). This shows the recognition of the customary forest is directly under the state forests, which are weak and most of the time dependent on the needs and aspirations of the state. Provisions set within the forestry law are less obliged to issue legal decisions on customary forests; entitlements for customary forest are hence overruled. Fortunately, indigenous peoples' struggle for their forestry rights gained momentum when the Constitutional Court made the decision in 2012 to revise the forestry law regarding which indigenous forest areas should be excluded from state forest. This decision set new challenges for the community to manage the forest and how they would benefit from the carbon market.

Issues concerning carbon tenure are mentioned in several governments' legal documents, which emphasizes carbon sequestration as one of the forms of ecosystem services. It should be noticed that those benefitting from carbon capture have the same stake as ones using environmental services. Through the abovementioned descriptions of state forest, carbon tenure prominently is restricted as a right to enterprise and does not guarantee ownership. However, the product of carbon sequestration may be sold just as other forest enterprises. There is also a provision that states the benefits from carbon enterprise need to be distributed both to the state and the nearby community. However, no provisions confer the process and means of distribution and the agency involved; the MoF or Ministry of Finance. The MoF has arrangements for customary forest schemes, but forest allocation maps show the forest to be bundled for other land uses and forest owed for timber extraction battles with customary forest assertions. Forest arrangements are disconnected from other forms of land-use provisions. Land under forest is organized by forestry law while land outside forest is regulated by basic agrarian law. Possible harmonization is through spatial law. Customary rights outside forest are relatively strong in recognition.

15.3.4 Conflicts in the Forest

The levels of conflict in the forestry sector have arisen following the New Order regime, whereby demands for natural resources share increased (Wulan and Yasmi, 2004). In particular, these conflicts highlighted past incidences of unfair benefit sharing with social significances in previous development models. These conflicts increased up to 11 doubles explicitly at 39%, 34%, and 27% occurrence at *Hutan Tanaman Industri* (industrial plantations) (HTI), conservation, and *Hak Pengusahaan Hutan* (forest concessions) (HPH) areas, respectively (Wulan and Yasmi, 2004). Boundary disputes, illegal timber felling, encroachments, degradation, and land-use changes are major categories for conflicts. More of the conflicts were in those areas that were overlapping protection forest with agricultural land (World Bank, 2000). There is also a series of conflicts between the central and regional governments on the issue of decentralization. Sequences of incidences have been reported between the local communities residing there and logging companies granted permission to undergo land-use changes. The government has not made complete efforts in solving forest conflicts. In the course of *reformasi* (post-Suharto era), an instantaneous devolution mechanism led to activating hidden conflicts while exciting new ones. Conflict resolutions were rendered through asking companies to pay compensation for logging concessions or industrial forestry plantations (HPH/HTI). This concept further angered the customary local people who had just lost their customary rights.

15.4 EQUITY, EFFECTIVENESS, EFFICIENCY: POLICY DIMENSIONS FOR REDD+

The main drivers of deforestation as discussed earlier are land-use change and illegal logging. Not all forms of deforestation may be avoided as they may aid in the livelihood of some people. On the other hand, in many cases, deforestation is the result of poor planning and inadequate law enforcement. This can be directed toward the overall policy, whereby some polices promote deforestations for economic incentives while others tend to reduce them.

15.4.1 Some Polices to Curb Deforestations

a. Environment laws: Indonesia's Law on Environment Protection and Management. if effectively implemented, may curb many problems of deforestation and degradations. This law demands the Ministry of Environment (MoE) to classify ecoregions along with local people residing there and their customs. This information forms the basis for formulating necessary environmental plans for climate change mitigation and adaption. The government is endowed to carry out environmental impact assessment (EIA) prior to implementing long-term development plans to ensure sustainable development planning. This process forms the major criterion for allocating business permits. This package is anticipated to have positive impact on REDD+ interventions. With the addition of public participation in the EIA process, REDD+ is likely to be effective.

b. Spatial planning: Several spatial planning policies demoralize deforestation and forest degradation. During spatial planning, public participation is mandatory to increase public interest. It sets out disciplinary actions for permit issuance that infringe rules. If this law is applied effectively, this can foresee successful REDD+ monitoring.

c. Miscellaneous polices: As soon as the Indonesian government provided the district government the autonomy to permitting business licenses, indiscriminate permits were issued without due considerations to social and environmental consequences. As such, the government withdrew such authority from the district government and this move has reduced deforestation in many ways.

As discussed earlier, polices promoting forest degradations include

a. **Land use polices:** This law triggers deforestations simultaneously, because spatial plans regulate forest borders, and in the current circumstance this is more favorable to sectors beyond forestry.

b. **Land tenure:** Land tenure is often confused with basic agrarian and forestry law and has conflicting provisions with adverse forest conditions. Heated arguments on who should own and manage land have aroused conflicts and controversies. Clarity and recognition over customary rights has abandoned forest monitoring initiatives with the present population increase and the tendency toward land and natural resources acquisitions that has reduced further forest damage. This potentially undermines successful REDD+ program interventions and adheres attainments of equity through the lack customary communities' participation.

c. **Regional autonomy polices:** Decentralization and authority devolution to district and regional governments increases deforestation through the number of forestry industrial permits granted without considering socioecological consequences. On the background of conflicting ideologies between economy and emission reduction, REDD+ is likely to be operative if it reimburses management and entities for "opportunity cost" suffered due to "disappointment" of economic inducements for which REDD+ substitutes. Regional government is likely to device REDD+ if it is relatively advantageous economically.

15.4.2 Assessment of REDD+ From Equity, Effectiveness, and Efficiency Perspectives

There are policies that if applied as envisioned create a favorable environment for REDD+. Activities listed under REDD+ strategy to curb deforestation and forest degradation also target social equity. Nevertheless, appropriate execution may prove intimidating as the development of economy under natural resources exploitation has multiple intertwined interests. The government's focus is on education and health, as evident through its stances on both short- and long-term plans. But development issues are linked with natural resources. REDD+ is likely to be effective from acceptable stakeholder support to the level it contributes to livelihoods and the economy. Negative waves created by weighty state dependence on natural resources are heaped by disaster to accumulate processes and activities. The segmented approach of issuing permits differently by different sectorial authorities, especially for mining and estate plantations, and weak monitoring by the central government violates sustainability. Also segmented are the EIA approaches. Even if archetypes of natural resources exist on paper, huge practice is foreseen. These segmented approaches in natural resources will destabilize efficacy of REDD+. The MoF with no stake in these segmented management scenarios cannot control permit issuance even at its early stages. With REDD+ interventions, irregularities are to be prevented though ensuring that all sectorial governments undergo the same process further strengthened by alerting policy makers for sustainable development. Bureaucratic reformation may reduce ambiguities on regulatory frameworks and monitoring systems. However, under the current administration, the Ministry of Environment and the Ministry of Forestry has been merged into the Ministry of Environment and Forestry. This merger is expected to synchronize environment and forestry issues management including REDD++ and the National REDD+ Agency, which has been incorporated under this new ministry.

Intricacies may be attributed to regional autonomy and associated politics. A problem has arisen because together the regional and central governments foresee protecting their own welfares without acknowledging passable public scrutiny and limitations on civil society's influence to engage in effective monitoring. Interpretation of forest and forest area is the result of such complexity followed by a series of extinguishing power and authority. There exists no criterion for performance evaluation of regional and central governments in forest management. This may have implications for REDD+ in conflicting priorities, power struggles, unclear authority, and effective transparent accountability mechanisms. The government is also weak regarding information management, effective participation, and fair justice. Attempts to enhance legal enforcement have yet to show benefits. Disappointment in detailing the "rule of law" is likely to make REDD+ susceptible to corruption threats predominant at all government levels. Such a situation of distrust will compromise REDD+ effectiveness, equity, and efficiency. With REDD+ in place, monitoring and reporting needs to be carried out by central administration while independent institutions conduct verifications. The dovernment has initiated forest inventories at the national level and built a carbon accounting system but lacks technical capacity.

The land tenure issue needs immediate attention because conflicts have been sparked between forest dwellers and customary people that affect forest conditions. The overall root cause surrounds ambiguities for customary rights recognition and claims; though recognition by law is practically not applicable. To fulfill their needs, these people are bound to disrespect laws and undertake hasty forest products.

15.5 CONCLUSIONS

This paper reviewed diverse dilemmas evolving in forest degradation and deforestation on the verge of implementing REDD+ in Indonesia. Several circumstantial encounters need to be addressed. The major difficulty in controlling deforestation is due to the fact that it is linked with the national economy. This dependency has changed the economy and landscape affecting REDD+ through a number of convolutions; explicitly, poor governance, power struggles between central and regional governments, immature spatial planning process, and gaps in policy and practice in customary rights and tenure arrangements. However, the government has made attempts to address many of these challenges through law promulgation and revisions.

Forestry law assigns areas under forest as "forest areas," but laws again confer the need for a list to establish "forest areas." This has led to multiple interpretations and less forest areas have been listed to date. Though the expanses are every so often under law, much of these are populated by indigenous locals demanding customary rights, and been been distributed for large development undertakings, explicitly for palm and timber plantations. These uncertainties on tenure arrangement are counterproductive in terms of SFM under REDD+. Although REDD+ has supplemented carbon value to the forest, it also has also resulted in contested privileges by a crowd of actors. This shows the need for improved land-use planning. The forestry sector passes under specific and sectorial regulation, which has led to a number of contradictions and inconsistencies and overall uncertainty about policy adherence by multiple interpretations; made worse by no coordination among pertinent institutions. This presents important challenges as some sectorial regulations promote deforestations.

Resourcefulness to lessen illegal logging should be a major segment in any emission reduction approach. REDD+ has latent appreciation to curb this logging through monetary incentives promoting compliance with existing legal provisions and negotiating changes in governance reforms. REDD+ activation may be concluded either through funds-based or market-based provisions, but this necessitates consideration in matters of reliability, traceability, and responsiveness to collective and governance safeguards, with all actions to be laid open to liberated verification. Harnessing modules from forest restructurings in miscellaneous countries will propagate prospects for retaining data sets, information allotment among ministerial actors, confirming distinct mandates for recognition, and overall standard situating.

References

Abood, S.A., Lee, J.S.H., Burivalova, Z., Garcia-Ulloa, J., Koh, L.P., 2014. Relative contributions of the logging, fiber, oil palm, and mining industries to forest loss in Indonesia. Conserv. Lett. http://dx.doi.org/10.1111/conl.12103.

Dermawan, A., Petkova, E., Sinaga, A., Mumu Muhajir, M., Indriatmoko, Y., 2011. Preventing the Risk of Corruption in REDD+ in Indonesia. United Nations Office on Drugs and Crime and Center for International Forestry Research, Jakarta and Bogor, Indonesia.

Galudra, G., Sirait, M., 2006. In: The Unfinished Debate: Socio-Legal and Science Discourses on Forest Land-Use and Tenure Policy in 20th Century Indonesia Paper presented at IASCP 2006 Conference. Retrieved from: http://www.iascp.org/bali/papers.html (retrieved 01.06.06).

Indrarto, G.B., Murharjanti, P., Khatarina, J., Pulungan, I., Ivalerina, F., Rahman, J., Prana, M.N., Resosudarmo, I.A.P., Muharrom, E., 2012. The Context of REDD+ in Indonesia: Drivers, Agents and Institutions. CIFOR, Bogor, Indonesia.

ITS Global, 2011. The Economic Contribution of Indonesia's Forest-Based Industries. ITS Global, Melbourne.

Mahdi, Shivakoti, G.P., Yonariza, 2015. Assessing Indonesian commitments and progress on emission reduction from forestry sector. In: Bajracharya, R.M., Sitaula, B.K., Sharma, S., Shrestha, H.L. (Eds.), Proceedings of the International Conference on Forests, Soils and Rural Livelihoods in a Changing Climate, Kathmandu, Nepal, p. 368.

Phelps, J., Webb, E.L., Agrawal, A., 2010. Does REDD+ threaten to recentralize forest governance? Science 328 (2010). Retrieved from: www.sciencemag.org.

Rusli, Y., 2011. Companionship on carbon in the region: basic idea and desire. In: Paper presented at the Asian Forum on Carbon Update 2011. DNPI and CENSUS, Hokkaido University, Bandung, Indonesia, 14 March 2011.

Safitri, M.A., 2010. Forest Tenure in Indonesia: The Socio-Legal Challenges of Securing Communities' Rights. Leiden University, Leiden, The Netherlands.

Soehartono, T., Mardiastuti, A., 2002. CITES Implementation in Indonesia. Nagao Natural Environment Foundation, Jakarta, Indonesia.

Sumardjono, M.S.W., 2008. Tanah dalam perspektif hak ekonomi, sosial, dan budaya. Penerbit Buku Kompas, Jakarta, Indonesia.

Sunderlin, W.D., Resosudarmo, I.A.P., Rianto, E., Angelsen, A., 2000. The effect of the Indonesian economic crisis on small farmers and natural forest cover in the outer islands. In: CIFOR Occasional Paper 29. CIFOR, Bogor, Indonesia. http://www.cifor.org/publications/pdf_files/OccPapers/OP-28%28E%29.pdf.

Verchot, L.V., Petkova, E., Obidzinski, K., Atmadja, S., Yuliani, E.L., Dermawan, A., Murdiyarso, D., Amira, S., 2010. Reducing Forestry Emissions in Indonesia. CIFOR, Bogor, Indonesia.

Wardojo, W., Masripatin, N., 2002. Trends in Indonesian Forestry Policy. Policy Trend Report. Ministry of Forestry, Indonesia. Retrieved from http://pub.iges.or.jp/modules/envirolib/upload/371/attach/06_Indonesia3.pdf.

World Bank, 2000. Indonesia: the challenges of World Bank involvement in forests. Evaluation country case study series. http://lnweb90.worldbank.org/oed/oeddoclib.nsf/b57456d58 aba40e585256ad400736404/749c3a7fe1d679c98525697000785b5a/$FILE/IndForCS.pdf.

Wulan, Y.C., Yasmi, Y., 2004. Analisa konflik:sektor kehutanan di Indonesia 1997–2003 [Annalysis of Forestry Sector Conflict in Indonesia 1997–2003]. CIFOR, Bogor, Indonesia.

Yonariza, Shivakoti, G.P., Mahdi, 2015. The gap between policy and practice in Indonesia forest rehabilitation. In: Bajracharya, R.M., Sitaula, B.K., Sharma, S., Shrestha, H.L. (Eds.), Proceedings of the International Conference on Forests, Soils and Rural Livelihoods in a Changing Climate, Kathmandu, Nepal, p. 368.

Toward an Effective Management of Dynamic Natural Resources

R. Ullah*, Yonariza†, U. Pradhan‡

*The University of Agriculture, Peshawar, Pakistan †Andalas University, Padang, Indonesia
‡World Agro-Forestry Center, Bogor, Indonesia

16.1 INTRODUCTION

Indonesia is a biodiversity and natural resource risk country covering around 191 million ha of land with a population of 270 million. A major proportion of the population of the country is engaged in the forestry and agricultural sector to earn their livings. The natural resources base plays an important role in providing livelihoods for the majority of people in Indonesia in general and in Sumatra in particular. The purpose of this chapter is to summarize the issues related to natural resources in Sumatra and the key findings of the case studies discussed in earlier chapters. The first section of the chapter presents a summary of the natural resources and related issues in Sumatra. The dependence of the local population on these natural resources for their livelihoods is summarized in section two, while the third section presents community participation and collective action (comanagement) in sustainable use of these natural resources. The role of polices and institutions in natural resources management is provided in the fourth section of the chapter. Recommendations, based on the findings of the case studies discussed in the above chapters, are provided at the end.

16.2 NATURAL RESOURCES AND RELATED ISSUES IN SUMATRA INDONESIA

Degradation of the natural resources base is on the rise in Indonesia and the context of natural resources management has dramatically changed, particularly during the last decade. The natural resources base of Indonesia is deteriorating mainly due to population pressure along with deforestation and uncontrolled soil erosion. Issues related to water quality and quantity are at the forefront and are major concerns of the government of Indonesia, donors,

221

and the public alike. Some of the issues discussed in the earlier chapters related to forests, land, water resources, protected areas, and biodiversity are summarized below.

16.2.1 Forests

Forests play a crucial role in sustaining water supplies, protecting the soils of important watersheds, and minimizing the effects of catastrophic floods and landslides. The idea of forest management is to recognize its ecological, social, and economic functions. The *Tata Guna Hutan Kesepakatan* (forest land-use policy) (TGHK) in 1999 categorized forests into six distinct categories: (i) preservation forest, (ii) protected forest, (iii) limited production forest, (iv) conversion production forest, (v) production forest, and (vi) other use forest. Under Article 6.1 of Law No. 41/1999 there are three main forest functions: conservation, protection, and production. Production forest has the main function of producing forest products; protection forest on the other hand has the main function of protecting life-supporting systems for hydrology, preventing floods, controlling erosion, preventing sea water intrusion, and maintaining soil fertility; and conservation forests enclose an area with specific characteristics having the main function of preserving plant and animal diversity and its ecosystem. Protection forest and conservation forest fall under the category of protected area.

Riparian forests and watersheds, if managed properly, can yield significant environmental services including high quality freshwater supplies. Forests play an important role in providing economic benefits, including employment and income for the population living around it. However, rapid deforestation and illegal logging have negative impacts on the livelihoods of the local communities, particularly those tied to forest products, and the environment as well. The local communities in Sangir, West Sumatra, negatively affect the sustainability of the forests in three different ways: (i) cultivating land inside the forests, (ii) mining activities, and (iii) harvesting forest products, particularly extraction of timber. Gold and tin mines have existed in the forest of West Sumatra since the colonial period. The Dutch gold mining companies extracting the minerals have already departed, as there is only a small portion of the minerals left in the area. The local people extract the remaining gold with traditional mining tools causing forestland degradation.

Land-use change and illegal logging are the leading causes of deforestation in Indonesia. Deforestation may not be avoided fully as it may aid in the livelihoods of local people. Poor planning and inadequate law enforcement also result in deforestation. Deforestation is linked to Indonesian finance and also shaped the political economy coupled with the institutional landscape, while affecting the forestry sector overall, explicitly REDD+. The government has realized this complexity, which is prominent through a number of reforms in national and sectorial polices, mostly trying to find a balance between allocation forest cover and agricultural land. However, the actual implementation of policy issued by the government is significantly low, creating a huge gap between policy and practice. Specifically with REDD+ implementation, Indonesia's national polices on emissions reduction are not included or not becoming mainstream policy yet in long-term national plans. This intricacy of trying to balance economy and reduce deforestation in the face of REDD+ creates a situation of complexity whereby the needs and problems of forest-dependent people may be lost. This is because polices are ad hoc to pursue space in the international REDD+ arena and furthermore to impose mandatory actions against local forest-dependent people to use and manage the resources to implement decisions made globally.

Deforestation causes a decrease of around 1.6 million ha of forest per year in Indonesia. Illegal logging, particularly by local communities, is not a new phenomenon in Indonesia. Illegal timber harvesting from protected areas poses a threat to tropical forests, and is an issue that has been attracting the attention of the international community because it is believed to cause environmental damage, promote corruption, increase poverty and social conflict, and can have variable but usually deleterious impacts on biodiversity and other globally important environmental services. Illegal logging is a complex problem and there are multiple causes for its existence in Indonesia. Some of the key causes of illegal logging include the need for income, whether other villagers (and nonvillagers) are already illegally logging, the recognition of the loss of community control over traditional forest areas, overcapacity of the wood processing industry, domestic and international demand for illegal timber, systemic corruption, rent-seeking behavior, rapid decentralization, growing unrest with the status quo, and poor law enforcement. Development policy that seeks to provide livelihood alternatives to timber-felling households could reduce dependence on timber and contribute to forest conservation in the nature reserve.

16.2.2 Land Resources

Land resources are capable of producing distinctive goods and services and make up the spatial plane on which most human activity takes place. In Indonesia, local people mostly depend on natural resources (mainly forest resources and agriculture) to earn their livelihoods. Highly dependent on those two sectors, the people realize the importance of proper and sustainable management of land resources. Land-use changes usually result from a greatly increasing population, economic growth, and physical development, all of which increase pressure on space and land resources. Along with population pressure, deforestation, uncontrolled soil erosion, the impact of opening land, and shifting cultivation are leading causes of land resources deterioration. The increasing pressure on land resources in turn increases conflicts and the loss of capacity of resources to maintain their functions. Population growth also raises land demand for settlement and public infrastructure. Converting the agricultural land into settlement area will reduce total rice paddy production of the area and will result in food insecurity as rice is a staple food for a majority of Indonesian citizens. Land-use changes at the "source" or upland areas of watersheds or river basins have effects on the users of the water downstream as it can influence the quantity, reliability, and quality of water downstream. Local communities have the rights to extract the resources on community-based property land and make optimal use of the land. The rights are transferred hereditarily from generation to generation without reducing the rights from the former to the future generation. The transfer of these utilization rights from the local community to the investors also impacts the local community and may cause the overexploitation of the natural resources base of the area.

16.2.3 Water Resources

The Barisan I Nature Reserve, covering an area of 74,000 ha in western part of West Sumatra Province, has the environmental function of maintaining water conditions in the catchments of several rivers that supply water to Singkarak Lake where a 154 MG hydroelectric power plant operates. These rivers also supply water to a number of small-scale irrigation systems surrounding the forest reserve. Thus it has an important role in hydrological regulation and is an important source of water for rural irrigation and nearby cities.

Water resources and related ecosystems provide and sustain all people, and are under threat from pollution, unsustainable use, land-use changes, climate change, and many other forces. Watershed and river basin management has generally achieved only partial success, largely due to the fact that biophysical factors have been emphasized at the expense of socioeconomic concerns and the fact that hydrologic boundaries are not congruent with political boundaries. Many efforts related to provision of clean drinking water and sanitation have been implemented by the Indonesian government. However, the performance of the water supply and sanitation sector in Indonesia is estimated to be lower compared to other countries in Southeast Asia. Around one-third of rural communities in Indonesia do not have access to clean water and two-thirds of the communities do not get access to sanitation. Lack of clean drinking water and improper sanitation causes widespread waterborne disease and diarrhea. In the Kuranji River Basin, West Sumatra, water quality issues related to surface waters are now being correlated to both hydrological characteristics and terrestrial biogeochemical processes, including land-use change and other basinwide anthropogenic issues. Land use affects pollutant yield, both in terms of the type of pollutant generated by a particular land use and in terms of total mass of pollutant released. Conflicts between *nagari*s and the local government over water and irrigation management also intensify the already worse situation. *Nagari*s claim that, based on a customary rule that recognizes the property rights of indigenous people, the irrigation canals within their territory are owned by them. This claim, however, is not formally recognized in regulations.

16.2.4 Protected Areas and Biodiversity

The Barisan I Nature Reserve in western part of West Sumatra Province is classified as a category VI protected area (PA), which "should contain predominantly unmodified natural systems, managed to ensure long term protection and maintenance of biological diversity, while providing at the same time a sustainable flow of natural products and services to meet community needs." The nature reserve is surrounded by densely populated villages in Solok, Padang Pariaman, Tanah Datar, and Padang districts. Eleven subdistricts and 23 *nagari*s are immediately adjacent to the nature reserve; irrigated rice farming is the main livelihood activity in these villages, and residents directly or indirectly use nature reserve products for their livelihoods. Forest types are non-*Dipterocarpaceae* lowland, submontane, montane, and subalpine. Some commercially valuable timber is found in all forest types, so timber cutting can occur at any altitude. The Barisan I Nature Reserve is a long-established protected area that dates back to Dutch colonial times in the early 20th century. The nature reserve, consisting of conservation and protection zones, plays a crucial role in ecosystem maintenance, particularly biodiversity conservation and water supply. Harvesting of timber is not allowed in both conservation and protection zones; however, local people can harvest nontimber forest products (NTFPs) in the protection zone.

Solok District contains several PAs, and the stakeholders have made a systematic effort to maintain these PAs. The district government of Solok set up Community Forest Guarding Units (CFGUs) in four *nagari*s, along the Barisan I Nature Reserve, consisting of the head of the *nagari*, chief of the youth wing, chief of the *adat* council, and the respective subvillage heads who are assigned to guard the forest. However, while some *nagari*s have taken initiatives to safeguard the protected areas in their vicinities, others have not been so successful and hence the forest conditions have varied across the *nagari*s.

In Indonesia, timber harvesting activities are commonplace in many PAs, especially since the collapse of a strong authoritarian government at the end of the 1990s. Timber harvesting in PAs involve multiple stakeholders including local people, logging companies, military personal, and forestry officials. The availability of valuable timber in commercial volumes, political change and bureaucratic confusion during transition in forest management policy, and lack of appreciation of PAs value causes illegal logging and other forest crimes in the PAs.

16.3 NATURAL RESOURCES AND LIVELIHOODS

The livelihood assets that supported livelihood strategies are natural capital, social capital, human capital, financial capital, physical capital, and religious capital. The access to capital assets is affected by the restoration of *nagari* rule in two ways. Local people occupy more land and extract more NTFPs from the *nagari* forest due to the uncertainty concerning rules of natural resources management during early stages of restoring the *nagari*. This uncertainty also discouraged local people from participating in *nagari* activities. Second, conflict over resource utilization causes insecurity of access to capital assets. Fishing, forest resources extraction, migration, and outside employment are the major activities of people from the low-income group, while people from the middle- and high-income groups tend to increase their livelihood security by intensification of agriculture and by investment in trading activities and small rural industry.

Fishing, harvesting the fish from the coastal water, is the main livelihood activity in terms of time allocation of the local community around the coastal areas of West Sumatra. However, fishermen whose livelihoods rely on less controlled and unpredictable fishery resources tend to experience higher uncertainty compared to other livelihood activities. Facing this uncertainty, the fishermen in Nagari Sungai Pisang tend to adopt (i) utilizing the kindness of nature, (ii) alternative productive activities, (iii) forecasting for unexpected future events, (iv) maintaining or enhancing social relationships, and (v) risk-spreading mechanisms and productivity enhancement. With these five livelihood strategies, the fishermen strived to secure their livelihoods.

Analysis of livelihood strategies revealed three major strategy elements: migration, job diversification, and agricultural intensification. Some households chose a combination of two or all three of these elements. People from different income groups differ with respect to which element they prefer. The tendency of migration, extraction of forest resources, and working in unskilled low-paying jobs and earning more income from remittances is on rise in the low-income households, while households of middle and high income tend to improve their livelihood security by occupational diversification toward more nonagricultural and off-farm activities. Though agriculture is still the most important source of income, there has been a decreasing trend for all household categories of deriving income from agriculture during the last 10 years.

The strategy of diversification in sources of livelihood includes cultivating rice paddy, providing services for tourism, making trawl based on demand, working as a ship maker, raising cattle, working as an agricultural wage labor, and performing nonagricultural jobs. After 1986 many local people whose main activity was previously fishing have turned to land-based livelihood activities. Cultivating rice paddy is considered as important strategy as it is capable of providing rice, the staple food of local people. As a result of this shift from fishing to paddy cultivation, the total land occupied by paddy field has significantly increased from 14 ha in 2004 to 73 ha in 2006. Other activities of extracting resources from nature's stock performed by the fishermen in Nagari Sungai Pisang were collecting firewood and clean water, which constitute important livelihoods for every family. Moreover, individual practical life skills helped in creating alternative livelihoods for the local people. Most fishermen in Nagari Sungai Pisang were capable of making trawl or fishing nets but only a few of them make these efforts sources of cash income.

Migration is another livelihood strategy practiced mostly by farmers in response to the insecurity and uncertainty of agricultural activities. The migrants leave the *nagari* to earn additional income by working in urban areas. There are two main factors that force the people with low incomes to find jobs outside. First, young people from the low-income group have little opportunity for higher education, and are generally not attracted to work in the agricultural sector. Second, employment growth within the *nagari* is not sufficient to absorb the growing labor force, which derives mostly from low-income households. Despite the fact that the low-income group has more access to land, the marginal growth in agricultural jobs is sufficient only to provide laborer jobs for the adult household members.

16.4 COMMUNITY PARTICIPATION AND COLLECTIVE ACTIONS IN NATURAL RESOURCES MANAGEMENT

Lack of incentives and the cost associated with participation in conservation efforts of natural resources limit the involvement and interest of regional and local governments to participate in conservation of the local natural resources base. These local and regional governments usually do not share national and international concerns over biodiversity conservation. In addition, the central government has been reluctant to involve local communities in PAs management, especially of conservation forest. However, if a clear incentive is available, local government is willing to share responsibility. After centralization changed to decentralization as the governmental system in Indonesia, the regional autonomy policy opens a possibility for comanagement of the natural resources. Comanagement, defined as "the sharing of power and responsibility between the government and local resource users" involves the central government, the local government, the commercial private sector, local communities, and civil societies. Comanagement along with the ongoing decentralization reform has created a better environment for effective management of community resources.

The central government, however, still holds strong control over the forest under Chapter I of Law No. 41/1999, especially in Article 4, which gives the central government the power to (i) regulate and organize all aspects of forest, forest areas, and forest products; (ii) determine the status of an area as a forest area or a nonforest area; and (iii) regulate and determine legal relations between humans and forest, and regulate legal actions concerning forestry. The role of local government in managing forest has been specified in the law. In implementing forest administration, the central government shall delegate part of its powers to local government. The roles of local government in managing forest areas within their jurisdiction, in this respect, shall be regulated by a central government regulation.

The government of Indonesia is planning to implement REDD+ without acknowledging the fact that forest management is a socioecological interaction that involves institutions, political pressures, and user actions that have come up through experiences and procedures. The accomplishments of REDD+ are incredible if natives are not involved in the implementation process. Although REDD+ is directed to benefit forest and native people, there is still uncertainty over poorly demarcated forest tenure and appropriate incentives agreed upon by the consensus building process.

Evaluation of the participatory activities in the past water supply and sanitation interventions supported by aid agencies in Indonesia have poorly addressed the issues related to equity in access for the poor, particularly women. The position and roles of women in the social community tend to place women as the weak party in the community. Despite the fact that men and women spend equal time on productive and reproductive activities, men have higher access to and control over family resources compared to women. Sociocultural factors and religious beliefs play an important role in the uneven distribution of access to and control of family resources and decision-making power/authority in favor of men. The important domestic work of women is usually unrecognized and undervalued and the opportunities for education and training are generally biased toward men.

Some of the local communities in Sumatra have organized internal microsaving and microlending schemes in response to difficulties in getting credit from formal institutions. *Kaum Melayu* in Nagari Dilam, for example, manages savings and credit for its members, in particular, to cover those household expenditures that exceed regular income; for example, medical treatment and elaborate traditional ceremonies such as weddings. *Balambiak Hari* is another form of collective action among the fishermen in West Sumatra. The group consists of usually 10 people, who help each other in certain activities of paddy cultivation. The collective action reduces the financial support required for paddy cultivation and provides an alternative source of earnings for the fishermen in the area. The fishermen sometimes also share the catch, as abundant catch has not always occurred; therefore sharing small portions of their catch would enhance their probability to get a share whenever other fishermen got fortunate. The generosity of catch sharing, especially when the catch was very abundant, had been locally institutionalized. The sharing activity is also carried out in crops as well in the area where abundant crop is shared with the less fortunate local people.

The local people in Nagari Paru have also developed a group that is responsible for monitoring and implementation of *Rimbo Larangan* (forest reserve). It consists of local individuals who are aware of the forest conditions and the group is called *Kelompok Petani Peduli Hutan* (KPPH), which is legally registered under SK Walinagari Number 188.47/05/Kpts-WN-2003. Community forestry (rehabilitation, conservation, and reforestation) in Nagari Paninggahan of Solok District is another collective activity in the area initiated by the *Gerakan Nasional Rehabilitasi Hutan dan Lahan* (National Movement of Land and Forest Rehabilitation) (GNRHL) with funding from Japan International Forestry Promotion and Research Organization (JIFPRO) with the contribution of Wakanyaku Medical Institute Ltd, Japan. It attempts to rehabilitate the critical land for recovering the function and increasing the productivity of land through a variety of plants (timber and nontimber). It provides an opportunity to increase the community welfare and recover the environment. Being a part of local community groups also yield benefits for livelihood strategies, as it could provide livelihood assets (especially natural capital) by accessing natural resources in the lowland forest. However, the consequences of these collaborative activities were all of the members had to allocate and adjust their daily activities to the group's schedules and appointments. It is very important for the local individuals to maintain a good performance and commitment within the team activity.

To gain active participation of the local people in the conservation of natural resources in the area, the GNRHL also provided daily wages for each activity called *Hari Orang Kerja* (HOK). The financial incentives motivate the community to appropriately manage their land and forest resources to solve environmental and natural resources problems. Besides the profitability, a number of other factors would also influence the community's decision of practicing in the conservation, rehabilitation, and protection of the natural resources base of the area. These factors include land tenure, availability of credit, technical assistance, and access to infrastructure (eg, market and transportation). Provision of these incentives will encourage the communities in implementing the conservation and rehabilitation activities in terms of sustainable use of natural resources.

The *nagari* is a promising local institution for natural resources management because decision making come from the local people. However, due to the fact that the decentralization act has been enacted only recently, the *nagari* is still not sufficiently capable to tackle all natural resources management problems. A number of conflicts over natural resources management issues have emerged within a *nagari*, between *nagaris*, and between *nagaris* and the local government. The policies and institutions need to strengthen the role of local people and local government in managing the natural resources base in the area.

16.5 POLICIES AND INSTITUTIONS FOR NATURAL RESOURCES MANAGEMENT

The growing concern over environmental problems caused by the degradation of the natural resources base in Indonesia as well as in Sumatra has led the government to develop and encourage several rehabilitation and conservation activities. The change of governmental systems from centralization to decentralization in Indonesia has led to the development of local regulations on the implementation of an autonomy system. Decentralization has been implemented in Indonesia since 2000. In response to this policy, West Sumatra Province has formulated a provincial regulation to give a legal basis for restoring the *nagari* institution and replacing the *desa* system of village administration, which used to be a common framework throughout Indonesia during Soeharto era. Through Act No. 22 of 1999 and Government Regulation No. 9 of 2000 about the village government system, which has been issued by the West Sumatra provincial government explains that districts, and village governments have autonomy to manage their own resources and to develop their own regulations in managing the resources. This wider autonomy for the village leader could be seen as an important driving force and to provide an opportunity to local initiatives for sustainable management of natural resources through their own decisions, based on their own aspirations.

The government of Indonesia, in 1999, issued Forestry Law No. 41/1999 replacing the old one. The new law provides a wider scope for involving all stakeholders, including local people and their institutions to manage their customary forest and acknowledges local customary laws on ecological and social aspects of forest management. Along with the restoration of the *nagari*, this law provides an opportunity for the people of West Sumatra to write and formalize their own customary laws regarding forestland and forest resources. Nagari Sungai Kamuyang of 50 Kota District, for instance, issued a *nagari* regulation in 2003 regarding the utilization of *nagari*-owned land, including forest resources. Understanding the importance of maintaining and increasing forest cover in the *nagari*, the government of Paninggahan also developed a number of initiatives (programs) for reforestation and rehabilitation and formed a collaboration with national and international stakeholders to build strong rehabilitation and reforestation projects, particularly for degraded land and forest area. Social forestry is another program covering all forest department programs aiming at decreasing forest fires and illegal logging and improving rehabilitation of critical land, structures forest industries, and push forest decentralization.

To be inclined with the process, the government has finalized guidelines to achieve the emissions commitment through reforestation and plantation under community forest and development of the *Hutan Tanaman Rakyat* People Plantation Forest (HTR). To propose justifiable community-based sustainable plantation forest management, HTR was familiarized by the state and REDD+ involvement with HTR is likely to fight to achieve anticipated outcomes. Indonesia has a huge span of tropical forest and has an immense role in accordance with international forums. The United Nations Forum on Forests (UNFF) is one of the international conventions that Indonesia is assigned to. With this agreement, governments have introduced priority policies in forestry, reducing illegal logging and forest fires, rehabilitating deforested land, and finally decentralized management. Surrounded by this arrangement, the social forestry approach was introduced to ensure sustainable forest and community prosperity together. Indonesia has also sanctioned the Convention on Biological Diversity (CBD) for achieving sustainable protection of biodiversity through a National Biodiversity Strategy and Action Plan (NBSAP).

Furthermore, in 2004, the government of Indonesia also issued new Law No. 7/2004 on water resources, replacing Law No. 11/1974. The new law recognizes responsibilities of provincial and district governments as well as of farmers in water management and also invites participation from private enterprises to manage and supply drinking water. The new law also influences household access to capital assets, including irrigation infrastructure and their participation in water management institutions. In addition, the new law emphasizes ensuring water resource conservation and protection. In relation to this new law, some *nagari*s in West Sumatra have formulated a *nagari* regulation on water resource management within their territory. Nagari Kinari, for example, enacted a regulation on irrigation management in 2004, which claims that irrigation canals within its territory are owned and should be managed by the *nagari*. The Indonesian government, with the financial support from the Asian Development Bank (ADB) has also launched some irrigation projects since 2001 to maintain irrigation canals and to enlarge irrigation coverage.

The government has provided subsidies to help poor households who face difficulties to gain access to some capital assets. Their access to social capital assets has, however, been reduced due to an increase in inequity that was caused mainly by the high-income group getting more benefits from market liberalization, and by poor households participating less in the newly reestablished institutions. However, their increased access to natural capital, as a result of new governance arrangements (ie, decentralization and restoration of *nagari* rule), stimulates intensive exploitation of the natural resources by the local people in response to economic shocks and cash needs.

16.6 RECOMMENDATIONS

Based on the findings of the case studies discussed in the above chapters the following recommendations are forwarded.

- Natural resources conservation will be more effective when the authority over it has been handed over to *nagari*s and other local institutions, after they have been empowered by defining their role in forest and water management and providing them with sufficient guidance and equipment.
- Provision of clear and certain laws from the central government is essential to guide local people to improve their environmental management institutional capabilities. This can be carried out through collaborative actions among different stakeholders/organizations including local government, nongovernmental organizations (NGOs), and universities. Financial support from international donor agencies is also needed.
- There is a need of immediate action to strengthen local institutions to conserve natural resources, improve livelihoods, as well as watershed sustainability, and to promote environmentally sound agricultural practices within the watershed. Issues related to water and land pollution resulting from higher use of agrochemical inputs and intensive soil tillage on steep slopes can be mitigated through the introduction/promotion of organic farming.
- The knowledge and awareness of environmental problems, its impacts, and the benefits of rehabilitation and conservation practices may help familiarize the people to become aware of the need for rehabilitation and conservation practices. Such knowledge and awareness can be spread through the use of print and electronic media and training programs.
- Local organizations should be organized by *adat*, religious figures, respected people, and older people in the village into a better management group. They have the capability to encourage people to participate in activities to build community empowerment. Women should also be encouraged to be active in community activities, the decision-making process, and in training held by the government or NGOs.
- Development policy that seeks to provide livelihood alternatives to timber-felling households could reduce local dependence on timber and at the same time contribute to conservation of the nature reserve. Additionally, forest conservation outcomes should improve if cooperation with local people in forest protection is developed, and if control over the resource is devolved to them.

Index

Note: Page numbers followed by *f* indicate figures and *t* indicate tables.

Printed in the United States
By Bookmasters